The Waters Above the Firmament

Dino Boccaletti

The Waters Above the Firmament

An Exemplary Case of Faith-Reason Conflict

Dino Boccaletti
Roma, Italy

ISBN 978-3-030-44170-8   ISBN 978-3-030-44168-5   (eBook)
https://doi.org/10.1007/978-3-030-44168-5

This Springer imprint is published by the registered company Springer Nature Switzerland AG
The registered company address is: Gewerbestrasse 11, 6330 Cham, Switzerland

To Ruben and Matilde

Preface

Usually, one writes a preface to a book intending to explain the reason or the circumstances which have inclined the author to write it. Often it is the author himself who writes the preface to his book. The present case is no exception. If one likes, one can instead consider exceptional the reason: the curiosity.

A couple of years before this book I am presenting, I wrote another one which had as its subject the history of the theories and experiences concerning the shape and size of the Earth, as its title says, from Homer to artificial satellites.[1] It was a subject regarding the history of science, or at any rate the history of the ideas which, even if very distantly, had to do with my past experience as a scholar of celestial mechanics.[2]

When studying the discussions on the shape of the Earth occurred in the Middle Ages (the spherical figure, already suggested since Aristotle's time, was not generally accepted), I stumbled across a subject whose existence I was unaware of until then: the waters above the firmament, which are mentioned in the verses 6-8 of the first chapter of the book of Genesis. Both the Fathers of the Church and the medieval authors, almost all members of religious orders, tackled the subject in the context of biblical exegeses concerning the book of Genesis (so the six days of creation, *Hexaemeron*) or, in the Scholastic period, in theological and philosophical works of a larger scope. The evident physical impossibility of a situation as the one set forth aroused my curiosity and led me to explore the discussions about that subject which had occurred from the early centuries of the Christian era to the late Renaissance with consequences in most recent times. So this book was born.

Since I was not aware of discussions about that subject outside the Church, as was the case for other controversial assertions in the Bible, I was also interested in grasping the reason for it. To this unexpressed question is due the subtitle of the book. In fact, it was apparently very strange that the lay culture had never brought into question an assertion so manifestly in contrast to physical reality. We shall return to the subject at the end of our research.

Of course, this is a book which moves in a very particular and bounded field. Its purpose is to narrate, and to comment upon, the conclusions which, in the course of around fifteen centuries, various authors have arrived at regarding the biblical narration of the second day of the creation. It is, so to speak, an inquiry (or a survey) based on "interviews" with a certain number of "experts" with a view to making known their opinions on a clearly defined problem. However, these experts are (were) not contemporary among themselves, as are experts in the present-day surveys. They speak to each other and to us across time.

As to the language I have used in this inquiry, I have tried to follow an approach which did not take many things for granted and therefore made them available to the non-experts. This is the reason for the biographical sketches that precede the "interviews"; these make no pretence of being thorough, but are intended only to introduce the "interviewee". As we all know, the interpretation of the Bible by exegetes of any extraction (believer or non), has given rise to

---

[1] Dino Boccaletti, *The Shape and Size of the Earth: A Historical Journey from Homer to Artificial Satellites* (Milan: Springer, 2019).

[2] See Dino Boccaletti and Giuseppe Pucacco, *Theory of Orbits*, 2 vols. (Heidelberg: Springer, 1996 and 1998).

thousands of works which have filled a staggering number of libraries in all the world. Therefore to venture an attempt to understand what those exegetes have concluded, even if only in regard to a few verses, can (rightly!) appear to be a daredevil idea. It is not possible to stand in as an exegete of the Bible, but one might aspire to be a critical reader of what he has written. Obviously, our reading will be a "lay" reading, which also has the aim of understanding the cosmogonic model that the idea of waters above the firmament refers to.

The book of Genesis was written a long time before the idea of a spherical Earth was suggested[3] and, as we shall see, the exegetes later found themselves in difficulty when attempting to reconcile the biblical writing with the results of the natural science; then the faith-science conflict was born, which has remained with us for almost two millennia.

Concluding, I have perhaps been too ambitious by tackling a problem which belongs to those traditionally treated by professional historians, but I hope that my journey over fifteen centuries of biblical exegeses may be of some avail both to non-specialized readers and to students.

A Guide to the Book

In the usual practice, in the context of Western culture, the term "science-faith conflict" alludes to cases in which a "new" scientific theory, or the result of an empirical research, brings into question what on the same subject is maintained in the Bible. Well-known emblematic cases are the Copernican theory of the planetary motions and the Darwinian theory of evolution; such cases are those in which a "new" scientific result contradicts (or, better, denies) the "literal" text of the Scriptures.

That which we shall deal with in this book is in some way a particular case. Here it is not a matter between the biblical text and a newly-discovered and indubitable scientific result; rather, the exegetes are trying to find an explanation for an assertion which seems to be in clear contrast to simple common sense. As we shall see in the course of our interviews, the exegetes are seeking an explanation which can satisfy the faithful (the exegetes wrote to guide the preachers and often preached themselves). We have to consider that for centuries the problem of the interpretation of Genesis 1:6-8 was debated within the Church, or in any case within the institutions delegated to preach the Christian religion. It is therefore certainly improper to talk over those verses with the way of thinking of a layman of our present day and to wonder why it has not turned out to be a faith-reason conflict in that regard. Indeed, at the beginning—that is, at the time of the first Christian exegetes—the situation was reversed, so to speak. It is the Christian exegete, with the uncompromising belief that what is written in Genesis is the "truth", who accuses the pre-existing Greek science of being wrong and the source of all possible "heresies" (for example, Pseudo-Hippolytus, Lactantius).

In fact, in this case, the problem is that of establishing the *true religion,* not the scientific "truth". Basil, for example, seems to accept what we could quickly call the Aristotelian vulgate with regard to the nature of the cosmos, but rejects some details which do not go with his literal interpretation of the biblical narration of the creation. However, Basil is not "against the science", it is simply that the problem does not stand in these terms yet. It will be Augustine who will know that the problem of a science-faith (indeed faith-science) conflict exists, even if in his writings it appears disguised by the question of how to intend the nature of an *ad litteram* interpretation.

Augustine is a philosopher and, on the strength of this, he realizes the profundity and the importance of the question. Obviously his discourse is more general than that of the existence

---

[3] D. Boccaletti, *The Shape and Size of the Earth,* op.cit., Chap. 1.

or not of the waters above the firmament we are interested in, nevertheless also this "local" problem will be "transferred" by Augustine to the philosophers of Scholastics.

Given the narrowness of the subject brought into question, the book is organized as a series of interviews which, still, are addressed to exegetes of Genesis belonging to different centuries (fifteen, to be precise). To introduce our interviews, which will begin with Origen, we shall hint at the cultural context of the first two centuries of the Christian era in which the exegetes of Genesis began to study. On the subject, the reader may find several books of eminent historians.

Our only task, introducing brief historical digressions, is to contextualize our "interviewees" in the cultural world in which they historically lived. The same will be done in the following.

A question we consider essential and which seems not always be perceived by the interviewees is: what is the cosmogonic model assumed by them who wrote Genesis? We shall call this model, only assumed but never explicitly expressed, the cosmogonic model of Moses and we shall discuss this in Chap. 4, in the conclusions. Our impression is that it has been a question tacitly removed by most of the exegetes, in the indestructible conviction that what is written in Genesis is the "truth".

Notice to the Reader

Most of the commentaries on Genesis quoted in this book were originally written in Latin. Of each quoted excerpt, the original Latin text is supplied in the footnote corresponding to the English version in the text of the book. When it has not been possible to find a "reliable" English version for various excerpts, I myself have provided a translation.

The equipment of Latin texts that appear in the footnotes might be considered exorbitant and perhaps useless, but we know, from personal experience, that many a time shades of meaning in a translation may lead to draw erroneous inferences. Therefore, at least for those who are familiar with Latin, *melius est abundare*.

Acknowledgments

The excellent editing by Kim Williams is gratefully acknowledged.

Rome, January 2020

CONTENTS

CONTENTS

Chapter 1
The Waters above the Firmament in the Early Centuries of the Christian Era

The subject we shall deal with appertains substantially to the interpretation of three verses at the beginning of the book of Genesis. They are:

> 6 *dixit quoque Deus fiat firmamentum in medio aquarum et dividat aquas ab aquis*
> 7 *et fecit Deus firmamentum divisitque aquas quae erant sub firmamento ab his quae erant super firmamentum et factum est ita*
> 8 *vocavitque Deus firmamentum caelum et factum est vespere et mane dies secundus.*[1]

We shall have to contextualize our problem historically, even if we shall only be able to graze, and superficially at that, subjects on which mountains of specialized bibliography of biblical exegeses exist. We shall start from the origins, that is, from the early centuries of the Christian era, when the diffusion of the Christian religion in the Roman Empire determined a new course of philosophy: "Every religion entails a set of *beliefs,* which are not the fruit of a research since they consist in the acceptance of a *revelation.* Religion is adhesion to a truth that man accepts by virtue of a superior *testimony.* Such is in fact Christianity. ... Therefore religion seems to exclude in its own principle the research, nay, to consist in the opposite attitude, of the acceptance of a truth testified to from above, independent of any research. Nevertheless, as soon as man wonders what the meaning of the revealed truth is and asks himself through which means he can truly understand it and make the flesh of his flesh and the blood of his blood, the need for research is reborn. ... Thus, the research is reborn from the religious itself because of the need of the religious man to draw as near as possible, to the revealed truth. ... From the Christian *religion* is thus born the Christian *philosophy*".[2]

Of course, the early Christian thinkers had at their disposal for their research the whole tradition of the Greek philosophy, from the pre-Socratics to the Stoics. But it fell to them to elaborate a doctrine which would *demonstrate* that the Christian religion was the point of arrival of classical Greek philosophy which, for its part, had always coexisted with the paganism.

The doctrinal "corpus" from which the Christians had to start was constituted, in addition to the Old Testament (i.e., the Bible), by the three Synoptic Gospels (Luke, Mark, Matthew), by the fourth Gospel (John), by the Epistles of St. Paul and by the Acts of Apostles. These texts had to be interpreted and a *correct* interpretation obtained. This interpretation, in its turn, would be become *religious* truth, accepted and professed by all faithful. This work was particularly arduous, especially in the early centuries, giving birth to internal conflicts within the Christian

---

[1] The version quoted above is that of the Vulgate (St. Jerome, end of the fourth century). The early Christians had at their disposal the ancient Greek and Hebrew versions. Since about the middle of the third century BC there was also the Greek version called "of the Seventy", which appears to have been carried out in Alexandria during the reign of Ptolemy Philadelphus (285–246 BC). This version is still used in the liturgy of the Greek Orthodox Church. Before the Vulgate, there existed other Latin versions considered unsatisfactory. Initially St. Augustine also used those versions.

[2] Nicola Abbagnano, *Storia della Filosofia* (Torino: Utet, 1993), vol. I, §128 (our Eng. translation).

© The Editor(s) (if applicable) and The Author(s), under exclusive license to Springer Nature Switzerland AG 2020
D. Boccaletti, *The Waters Above the Firmament,*
https://doi.org/10.1007/978-3-030-44168-5_1

communities and ideological battles with different philosophical currents. The Christian thinkers, from the very beginning, besides the elaboration of the religion, therefore also often had the tasks of preaching to the faithful the *correct* interpretation of the Holy Scriptures and of fighting the *heresies*.

In their activity some thinkers distinguished themselves as those who elaborated the bases of the Catholic Church; for this, they were later declared to be *Fathers of the Church* and the period of the history of philosophy regarding them of the *Patristics*. The real philosophical activity of the Fathers of the Church begins in the second century with the *Apologists,* who write in defense of Christianity by arguing, sometimes quite harshly, with Hebrews and pagans. There were Eastern Apologists (Justin, Tatian, Athenagoras, Theophilus of Antioch, and others) and, successively, Latin Apologists (Tertullian, Minucius Felix, Lactantius, and others), to name only the most renowned. Besides the works written "against" a particular interlocutor, also there were works directed towards the Christian faithful to help them identify the *heresies*. To the early period belong works which, while of dubious attribution, are nevertheless considered to be important in the development of the disputations against the heresies. One of these, as we shall see, is *The Refutation of all Heresies*, which was for a long time attributed to Hippolytus of Rome.

1.1 Irenaeus of Lyon and Hippolytus of Rome

Before beginning to deal with the interpretations which tackle the three famous verses we have quoted above, we cannot avoid talking about the works which handled the problem from an overall point of view, that is, the works that fought "all" the "misconceived" interpretations (the *heresies*). Perhaps the most indicative is the one we have already mentioned, *The Refutation of all Heresies*.[3] This work was preceded by one bearing a similar title by Irenaeus,[4] who (although this is not established unequivocally) appears to have been the master of Hippolytus, who in his turn, as we know, seems not to have been the author of *The Refutation of all Heresies*. However things may really stand (about the authors and the works of that period, the hypotheses of attribution have been changed often in the last two centuries), the work by Irenaeus was a fundamental forerunner.

Irenaeus was born in Smyrna in 130 and died in Lyon in 202. He was a pupil of Polycarpus, bishop of Smyrna and reputed pupil of the apostle John. He became bishop of Lyon in 177, following the death of the previous bishop, a martyr of the persecution under Marcus Aurelius. He was a strong opponent of gnosticism, which was the first search for a philosophy of Christianity, in which knowledge (*gnosis*) was held to be the condition for safety. Against this heresy, he wrote the work handed down with the Latin title *Adversus haereses*. That work, as a matter of fact, was written in Greek, but the extant unabridged text is that of a Latin translation of the fifth century; it seems that the original Greek title translates as *On the detection and overthrow of the so-called Gnosis*. It consists of five books, the first four of which are devoted to fight particular gnostic currents.

The battle against the gnostic currents is the principal effort of the Christian theology in this period and it is actually in this battle that it clarifies its definition and its principles. A principle

---

[3] See Pseudo-Ippolito, *Confutazione di tutte le eresie*, trans. and ed. Augusto Cosentino (Rome: Città Nuova, 2017). The editor, in the Introduction, gives a detailed account of the studies on the work and its attributions. The attribution to Hippolytus, as it can be seen from the title, is dropped.

[4] See Irenaeus of Lyon, *Against Heresies* (Beloved Publishing, 2015). See also Ireneo di Lione, *Contro le eresie*, ed. Augustro Cosentino (Rome: Città Nuova, 2009).

forcefully held by Irenaeus is that of "apostolic succession", according to which the apostles hand down their authority to their successors, the bishops, heads of the churches.

Now, we can talk about the work of Pseudo-Hippolytus, which can also be cited by the Latin title of *Refutatio* or the Greek title *Elenchos*.[5] This work is supposed to go back to the first decades of the third century (depending on scholars, the dates vary from before 222 to the decade 230–240). The heresies described in the *Refutatio* are not all classifiable as gnostic, while in the work of Irenaeus they were the only one considered. Further, in comparison to Irenaeus's work, besides exclusions, there are also additions, allowing scholars to have, in the aggregate, a sufficiently exhaustive picture of the gnostic sects.

The work, in its materiality, has suffered a rather hard history. In fact, in its original integrity, it was divided into ten books, but the second, the third and the first part of the fourth have been lost. It is thought that the first four books were all devoted to a succinct exposition of the doctrines of the Greek philosophers (Aristotle did the same thing, in his time, in the *Metaphysics*), with the aim of demonstrating that all the Christian heresies had their origins therein. This introductory part about Greek philosophy was known by the Greek title *Phylosophumena* and was for a long time attributed to Origen (of whom we shall speak below). Its rejoining with the fifth through tenth books took place in the mid-nineteenth century, when new manuscripts were found. The real refutation of the heresies is the second part of the work. Let us see how the author justifies his exposition of the Greek philosophy. He says in the Proem, "We must reject no myth touted by the Greeks. Their inconsistent doctrines must be considered trustworthy on account of the excessive insanity of the heretics. Most people suppose that these heretics worship God due to their silence and the concealed nature of their secret mysteries. A long time ago, I presented their doctrines in a limited fashion, not exposing them in detail, but refuting them in a general way. I did not think it proper to bring their unspeakable mysteries into the light, so that, when I presented their opinions in enigmas (ashamed as I was to declare what is secret and expose them as godless), they might cease somehow from their irrational mind-set and unlawful endeavor. But since I see that they have not blushed before my leniency, nor taken into account the patience of God (though blasphemed by them) that they might repent out of shame or, by remaining intransigent, be judged in righteousness, I will proceed—forced as I am—to reveal their secret mysteries! … Yet no other will refute these teachings except the Holy Spirit handed on in the church. The apostles obtained this Spirit beforehand and shared it with orthodox believers. I, their successor, participate in the same grace of high priesthood and teaching. Accounted as a guardian of the church, my eyes do not sleep, nor do I keep secret the orthodox doctrine. Nevertheless, not even as I labor with my whole soul and body do I grow faint as I endeavor worthily to make worthy repayment to God my benefactor. Not even by such labor do I make a fair exchange! Still, I am not slack in those matters with which I have been entrusted. I fulfill the due measures of my time and generously share with all whatever the Holy Spirit provides. I preach without shame, bringing out into the open and dangerous material through an exposé but also whatever the Truth has received by the Father's grace and administered to human beings—these things I both record from conversation and bring as testimony from written reports".[6]

At last, in the Conclusion of the fourth book, he says, "It appears, then, that these speculations also have been sufficiently explained by us. But since I think that I have omitted no opinion found in this earthly and grovelling Wisdom, I perceive that the solicitude expended by us on these subjects has not been useless. For we observe that our discourse has been

---

[5] See the introcution by Augusto Cosentino in Pseudo-Ippolito, *Confutazione di tutte le eresie*, op. cit.

[6] M. David Litwa, ed. *Refutation of all Heresies* (Atlanta: SBL Press, 2015), pp. 3-7.

serviceable not only for a refutation of heresies, but also in reference to those who entertain these opinions. Now these, when they encounter the extreme care evinced by us, will even be struck with admiration of our earnestness, and will not despise our industry and condemn Christians as fools when they discern the opinions to which they themselves have stupidly accorded their belief".[7]

We have preferred to quote in full the above excerpts instead of condensing their content in order to allow the reader to get the idea both of the style used by the author and of his *vis polemica*.

The *Refutatio*, as we have seen, is a work which has the aim of forewarning Christians of falling into "heretical enunciations", both by indicating in the works of the Greek philosophers a possible origin of the heresies and listing and commenting on the currents of heretical thought underway at that time. In it, there is no explicit reference to the Bible, but several exegetic treatises are attributed to Hippolytus, the greater part of which gone lost or reduced into fragments. As regards the Bible, the treatises are extant on the Song of Songs and on the Book of Daniel; no commentary on the book of Genesis came out.

Instead, we can trace an excerpt, half paraphrase and half commentary, devoted to the verses of Genesis that we are interested in, in the work of one of the Apologists quoted above, Theophilus of Antioch. We know neither his date of birth nor birthplace, but it seems to have been verified that he was Syriac and died about 183–185 as bishop of Antioch. The work is *Apology to Autolycus*,[8] in three books, against Autolycus, defined by Theophilus as an idolater who scorns the Christians. In the second book, there are several chapters devoted to the six days of Genesis. *Inter alia*, Theophilus, who is the first Christian author who speaks about the Trinity of God, asserts that the three days of the creation of light are an image of the Trinity of God, of its Word, of its Wisdom. But let us see what he says about the waters.

In the twelfth chapter, he exalts the description of the work of the six days, also quoting the description of the creation of the world in Hesiod's *Cosmogony*. In the thirteenth chapter, he goes into detail of the creation of the world and interprets the famous verses. Theophilus says that God places above the firmament (the heaven we see, distinct from the other heaven unseen for us) half of the waters, which will serve for the rains, the showers and the dew destined for humanity's benefit. The other half is left on the earth for the rivers, the sources, the seas. The whole is properly distributed and adorned with grass, plants, etc. The argument about the supracelestial waters destined for the atmospheric phenomena, as we shall see, will also be repeated by the writers of the Middle Ages.

1.2 Origen (184–253)

Among the institutional structures of the Catholic Church at its origins, one must count the catechetic and theological school of Alexandria, which is said to have been founded, or refounded, by Pantaenus in 180. Pantaenus, of whom there is no relevant biographical note, was born in Sicily and died at Alexandria in 200. His activity in the school was continued by Clement of Alexandria (Athens? ca. 150–Caesarea ca. 215) and by his pupils Alexander of Jerusalem and Origen. Now we shall deal with Origen and his Homilies on Genesis.

Origen was undoubtedly the most important exponent of the Christian philosophy of the first three centuries and the first, among the Fathers of the Church, who provided an organic and complete doctrinal formulation, even if during the course of his life he had several clashes of

---

[7] Eng. trans. Alexander Roberts, James Donaldson and Arthur Cleveland Coxe, from *The Ante-Nicene Fathers: The Writings of the Fathers*, vol. V (New York: Cosimo Classics), pp. 41-42.

[8] See Theophilus of Antioch, *Ad Autolycum*, trans. Robert M. Grant (Oxford: Clarendon Press, 1970).

opinion with the hierarchy, which did not always accept his ideas. We are only interested in him as a commentator of the book of Genesis; indeed we shall limit ourselves to his first homily, focussing as always on the verses 6-8. In any case we supply some succinct notes, necessary for situating the subject we are interested in.

Origen was born, reputedly, at Alexandria around 185, according to the notes on his life in the work *Church History* by Eusebius of Caesarea (263–339).[9] He came from a Christian family (his father was martyrized in 202 during the persecutions of Septimius Severus) and educated to the knowledge of the Greek culture and of the Bible. He assumed the direction of the catechetic school of Alexandria, was appointed by the bishop Demetrius when still quite young (perhaps 18 years old, according to Eusebius, but later according to others). Very soon he found himself in opposition to Demetrius, perhaps because of the importance he attached to philosophy in the elucidation of the *truth* of the faith, and he withdrew to Caesarea of Palestine, near by his friend Teoctistus. In Caesarea he founded a theology school, as a successor to that of Alexandria, and met with great success.

In Caesarea, besides teaching, he also devoted himself to preaching and writing many works, the majority of which are lost. Among those extant, many in the Latin translation by Rufinus (345–510), there are commentaries to books of the Bible and the doctrinal works *De Principiis* and *Contra Celsum* (written to refute a writing of the Neoplatonist Celsus). Origen remained in Caesarea until his death, which happened because of the tortures suffered during the persecution under Decius (249–250).

A fundamental characteristic of the exegesis of the Holy Scriptures by Origen is the emphasis on the allegorical meaning. The allegorical interpretation of the Old Testament had already been suggested by the Hebrew philosopher Philo of Alexandria (ca. 250 BC-ca. 45 AD),who put forward an allegorical interpretation (in this case, we must say, of the Torah) through the use of the Hellenistic-Alexandrian culture. Whereas Philo had founded his allegorical interpretation on the *logos*, Origen founds his interpretation on the historical reality of Christ. According to him, it is Christ's coming which unveils the profound meaning of the Holy Scripture.

Obviously, it was not easy to explain the allegorical meaning to all audiences and in the homilies Origen often run into trouble since many of his faithful were simple souls and not members of the Hellenistic cultural milieu.[10] As an example, in the first homily on Genesis, which (of the sixteen remaining) is the one that appertains to the question we are interested in, Origen proposes an allegorical interpretation completely neglecting any cosmological hint. Origen comments the verse 7 in this way: "Although God had already previously made heaven, now he makes the firmament. For he made heaven first, about which he says, 'heaven is my throne'. But after that he makes the firmament, that is, the corporeal heaven. For every corporeal object is, without doubt, firm and solid; and it is this which 'divides the water which is above heaven from the water which is below heaven'. For since everything which God was to make would consist of spirit and body, for that reason heaven, that is, all spiritual substance upon which God rests as on a kind of throne or seat, is said to be made 'in the beginning' and before everything. But this heaven, that is, the firmament, is corporeal. And, therefore, that first heaven indeed, which we said is spiritual, is our mind, which is also itself spirit, that is, our spiritual

---

[9] Eusebius, *The Church History: A New Translation with Commentary*, trans. and ed. Paul L. Maer, (Kregel Academie, 1999), p. 207.

[10] On this, see "Omelie e Commentari esegetici-Origene e il suo pubblico" in: *Origene-Omelie sulla Genesi*, ed. Manlio Simonetti, Ital. trans. Maria Ignazia Danieli (Rome: Città Nuova, 2002), pp. 11-16. See also the English edition Origen, *Homilies on Genesis and Exodus*, trans. Ronald E. Heine (The Catholic University of America Press, 1982).

man which sees and perceives God. But that corporeal heaven, which is called the firmament, is our outer man which looks at things in a corporeal way.

"As, therefore, heaven is called the firmament because it divides between those waters which are above it and those which are below it, so also man, who has been placed in a body, will also himself be called heaven, that is, heavenly man," in the opinion of the apostle Paul who says: 'But our citizenship is in heaven' … Let each of you, therefore, be zealous to become a divider of that water which is above and that which is below. The purpose, of course, is that, attaining an understanding and participation in that spiritual water which is above the firmament one may draw forth 'from within himself rivers of living water springing up into life eternal' removed without doubt and separated from that water which is below, that is, the water of the abyss in which darkness is said to be, in which 'the prince of this world' and the adversary, 'the dragon and his angels' dwell, as was indicated above. Therefore, by participation in that celestial water which is said to be above the heavens, each of the faithful becomes heavenly".[11]

The allegory regarding the nature of heavens thus allows one to evade any cosmological interpretation more or less tied to the natural philosophies of that time. But it should be noted that the interpretation of the 'inferior waters' as the seat of the demons did not meet with a favorable response of the subsequent commentators. We shall find an attitude open to a more literal interpretation with reference to the cosmological model of that time in Basil more than a century later.

1.3 Lactantius (ca. 250–after 317)

In chronological succession, after Origen, we shall deal with one of the Latin apologists, perhaps the most renowned one. Lucius Caecilius Firmianus Lactantius was born in Roman Africa around 250, more than half a century after Origen. His family was pagan and, according to St. Jerome, he was a pupil of Arnobius at Sicca in western Africa close to Carthage. Because of his reputation as a rhetorician, he was summoned by Diocletian to teach Latin rhetoric at Nicomedia (the capital of ancient Bithynia, modern Izmit), where he moved around 290. In that period he was converted to Christianity and in 305 he was removed from his chair in consequence of the tetrarchic persecution. After a period in which he was in difficulty, he enjoyed new favor and was summoned by Constantine to become the tutor of his son Crispus.

---

[11] Homily I: *Cum jam antea Deus fecisset coelum, nunc firmamentum facit. Fecit enim coelum prius, de quo dicit: coelum mihi sedes. Post illud autem firmamentum facit, id est corporeum coelum. Omne enim corpus firmum est sine dubio et solidum; et hoc est quod dividit inter aquam, quae est super coelum, et aquam, quae est sub coelo. Cum enim omnia quae facturus erat Deus, ex spiritu constarent et corpore, ista de causa in principio et ante omnia coelum dicitur factum, id est omnis spiritalis substantia, super quam velut in throno quodam et sede Deus requiescit. Istud autem coelum, id est firmamentum, corporeum est. Et ideo illud quidem primum coelum, quod spiritale diximus, mens nostra est, quae et ipsa spiritus est, id est spiritalis homo noster qui videt ac perspicit Deum. Istud autem corporale coelum, quod firmamentum dicitur, exterior homo noster est, qui corporaliter intuetur. Sicut ergo firmamentum coelum appellatum est ex eo quod dividat inter eas aquas, quae sunt super ipsum, et eas, quae sub ipso sunt, ita et homo, qui in corpore positus est, si dividere potuerit et discernere quae sint aquae sunt superiores super firmamentum, et quae sint quae sunt sub firmamento, etiam ipse coelum, id est coelestis homo, appellabitur secundum apostoli Pauli sententiam dicentis:* nostra autem conversatio in coelis est … *Studeat ergo unusquisque vestrum divisor effici aquae eius quae est supra et quae est subtus, quo scilicet spiritalis aquae intellectum et participium capiens eius quae est supra firmamentum,* flumina de ventre suo educat aquae vivae, salientis in vitam aeternam, *segregatus sine dubio et separatus ab ea aqua quae subtus est, id est aqua abyssi, in qua tenebrae esse dicuntur, in qua princeps huius mundi et adversarius draco et angeli eius habitant, sicut superius indicatum est. Illius ergo aquae supernae participio, quae supra coelos esse dicitur, unusquisque fidelium coelestis efficitur* (Eng. trans. from Origen, *Homilies on Genesis and Exodus*, trans. Ronald E. Heine (Washington DC: Catholic University of Americal Press, 1982), pp. 49-50).

He carried out this job at Treves between 313 and 316, and there probably died after 317. His cultural education was tied to the classical culture, especially Neoplatonism, while in adult age he absorbed the Christian theological and doctrinal elements.

Whereas his works preceding the conversion are lost, several apologetic writings are extant: *De Opificio Dei* (303–304), *Divinae Institutiones* (304–313), *De ira Dei* (after 313), *De mortibus persecutorum* (316–321) and the symbolic poem *Carmen de ave Phoenice*, in which the myth of the phoenix that arises from his ashes gives a faint idea of the resurrection of Christ. The main work by Lactantius is unquestionably the *Divinae Institutiones*,[12] which constitutes, in the West, the first attempt to reduce the Christian doctrine to a system constructed in an organic and complete way. The work consists of seven books. The first three (*De falsa religione, De Origine erroris, De falsa sapientia*) refute the foundations of the pagan religion, while the remaining four (*De vera sapientia et religione, De iustitia, De vero cultu, De vita beata*) enunciate the Christian doctrine in a systematic way. Our attitude in scanning the chapters of this work is obviously not the same of those who are interested in looking for theological enunciations and comparing them with those of other Fathers of the Church but, more simply and "secularly", to look in them for possible passages where Lactantius suggests an interpretation of the cosmogony narrated in Genesis.

Book II (*De origine erroris*) hints at the subject in several points. Let us begin from the chapter V: "How much more correct it is, therefore, to disregard insensible and vain images and to turn the eyes to where is the seat, to where is the abode of the true God! He is the God who supports the earth, who has adorned the sky with gleaming stars, who has illuminated the sun—a most brilliant light for human affairs and a singular witness unto His one majesty—who has spread the seas about the lands, who has charged the rivers to flow with everlasting roll, and who 'did order the plains to stretch out and the valleys to subside, the woods to be covered with foliage and the rocky mountains to arise.'[13] Surely not Jupiter, he who was born one thousand seven hundred years ago, did all these things, but rather 'That maker of things, cause of a better world,'[14] who is called God, whose beginning since it cannot be comprehended ought not to be even sought. It is enough for man, unto full and perfect wisdom, if he understand that that maker is God. Of this understanding the highest power is this, that it look up to and honor the common parent of the human race and the maker of wonderful things. Whence do certain people of blunt and dull heart adore the elements, things which have been made and which lack sensibility, as though they were gods? It is because, when they gazed in wonder at the works of God, namely: the sky with its various lights, the earth with its plains and mountains, the sea and rivers, lakes, and springs, struck dumb with admiration at these things, and forgetting their very Maker whom they were not able to see, they began to venerate and worship His works. Nor could they ever understand how much greater and more wonderful is He who made those works out of nothing. Although they see that these things, following divine laws, serve the convenience and uses of man with perpetual necessity, they still think that they are gods. They who prefer His works to God, the most indulgent Father Himself, are ungrateful toward the divine benefits. But why is it strange if barbarians or uncultured men are in error, when even philosophers of the Stoic school are of the same error, since they hold that all the heavenly bodies which move ought to be counted in the number of the gods? ... And we prove, O philosophers, that you are not only uninstructed and impious, but even blind, foolish, and mad,

---

[12]See: *Divinarum Institutionum Liber I,II,....,* VII-MPL 006 and Lactantius, *The Divine Institutes-Books I-VII,* trans. Mary Francis Mc Donald, O.P. (The Catholic University of America Press, 1964; rpt 2008).

[13] Ovid, *Metamorphoses,* I.43-44.

[14] Ibid. I.79.

for you have overcome the ignorance of the unlettered with emptiness. Whereas they considered as gods the sun and moon, you think even the stars are gods."[15]

Besides Ovid's *Metamorphoses*, a little below Lactantius also quotes Cicero's *De natura deorum*, an indication that he thinks that his readers will know him; that is, he believes that the receivers of this work will be the Roman educated elite. Ovid's verses are even used to continue the paraphrase of the first verses of Genesis. One has the impression that Lactantius wants to embellish the concise and solemn narration of Genesis.

He continues to talk about the creation of the world in the tenth chapter, where he resumes his discussion of the nature and function of the elements, again quoting Ovid. He also discusses the Earth, listing the four cardinal points and assigning to God the East as his seat. From his discussion it clearly emerges that the Earth is considered a flat disc. As we shall have the means to discuss in the last part of the present book, Lactantius substantially gives a correct interpretation of what is written in Genesis. He goes back to the same theme in the twenty-fourth chapter of the third book (*on the Antipodes, the heavens and the stars*): "For men are always deceived in the same way. Since, led on by a likeness of the truth, they have taken hold of something false in the beginning, it is necessary for them to run into those errors which follow. Thus, they fall into many ridiculous things because those things which are congruent with false things must be false. Although they have put faith in the first, they do not perceive what sort are those which follow, but they defend them in every way, even though they ought to judge whether the first things are true or false from those that follow. What reason, therefore, led them to the antipodes? They saw the courses of the stars wandering into the decline or setting, and they saw the sun and moon always setting into the same direction and rising from the same direction. But when they did not perceive what machination directed their courses or how they returned from setting to rising, and since they thought that the sky itself was sloping into all directions (it is necessary for it to seem so on account of its immense expanse), they believed that the world was round like a pillar; and they thought that the heavens revolved from the motion of the stars; and, thus, the stars and the sun, when they set, were brought back to their rising by the very whirling of the world. And so they fabricated aerial orbs, as it were, according to the form of the world, and they adorned them with certain portent-like images which they said were stars. That followed this roundness of the sky, therefore, so that the earth

---

[15] Lactantius, *The divine Institutes*, op.cit., II, 5: *Quanto igitur rectius est, omissis insensibilibus et vanis, oculos eo tendere, ubi sedes, ubi habitatio est Dei veri; qui terram stabili firmitate suspendit; qui coelum distinxit astris fulgentibus; qui solem rebus humanis clarissimum, ac singulare lumen, in argumentum suae unicae majestatis accendit: terris autem maria circumfudit, flumina sempiterno lapsu fluere praecepit. "Jussit et extendi campos, subsidere valles / fronde tegi silvas, lapidosos surgere montes". Quae utique omnia non Jupiter fecit, qui ante annos mille septingentos natus; sed idem: "Ille opifex rerum, mundi melioris origo", qui vocatur Deus, cujus principiimi, quondam non potest comprehendi, ne quaeri quidem debet. Satis est nomini ad plenam perfectamque prudentiam, si Deum esse intelligat: cujus intelligentiae vis et summa haec est, ut suspiciat et honorificet communem parentem generis humani, et rerum mirabilium fabricatorem. Unde quidam hebetis obtusique cordis, elementa, quae et facta sunt et carent sensu, tamquam deos adorant. Qui cum Dei opera mirarentur, id est coelum cum variis luminibus, terram cum campis et montibus, maria cum fluminibus et stagnis et fontibus, earum rerum admiratione obstupefacti, et ipsius artificis obliti, quem videre non poterant, ejus opera venerali et colere coeperunt; nec umquam intelligere quiverunt, quanto major quantoque mirabilior sit, qui illa fecit ex nihilo. Quae cum videant divinis legibus obsequentia commodis atque usibus hominis perpetua necessitate famulari, tamen illa deos existimant esse; ingrati adversus beneficia divina, qui Deo et patii indulgentissimo sua sibi opera praetulerunt. Sed quid mirum, si aut barbari, aut imperiti homines errant? cum etiam philosophi Stoicae disciplinae in eadem sint opinione, ut omnia coelestia, quae moventur, in deorum numero habenda esse censeant; ... ac vos, o philosophi, non solum indoctos et impios, verum etiam caecos, ineptos delirosque probamus, qui ignorantiam imperitorum vanitate vicistis. lili enim solem et lunam, vos etiam sidera deos putatis* (Eng. trans. by John J. Savage).

might be enclosed in the depths of its center. But if this were so, that the earth itself also is like a globe, neither could it be possible that what was held enclosed by roundness should not be round. If, however, the earth were also round, it is necessary that it bear the same appearance into all the parts of the sky, that is, that it put up mountains, stretch forth plains, spread out seas. If this were so, then that last point also would follow, that there is no part of the earth which is not inhabited by man and other animals. Thus, the roundness of the heavens comes up against those hanging antipodes. But if you ask those who defend these portents how, then, all things do not fall into that lower part of the sky, they will answer that this is the nature of things, that weights be borne into the middle and that all things be connected at the center, just as we see the spokes in a wheel; things which are light, however, as clouds, smoke, and fire, are scattered from the center, so that they might seek the sky.

"I do not know what to say about those who, when once they have gone astray, constantly remain in their foolishness and defend their empty theses with empty prattling, except that sometimes I think that they philosophize for the sake of a joke, or that cleverly and knowingly they take up lies to defend them, so that they might, as it were, exercise or demonstrate their abilities in evil things. But I could prove by many arguments that it is in no way possible for the sky to be below the earth, except that this book must be concluded now and there still remain some points which are more necessary to the present task. And since to run through the errors of individuals is not within the scope of one book, let it be enough to have enumerated a few, from which it can be understood of what sort the others are."[16]

Whereas in the first passage (II, 5) Lactantius seems to hold a relatively relaxed tone in his exposition, in the second one (III, 24) lets himself go to excited expressions, more suitable for an address, leading William H. Stahl to consider him one of the most embittered among the

---

[16] Lactantius, *The divine Institutes*, op.cit., III.24: *Nam semper eodem modo falluntur. Cum enim falsum aliquid in principio sumpserint, veri similitudine inducti, necesse est eos in ea, quae consequuntur, incurrere. Sic incidunt in multa ridicula; quia necesse est falsa esse, quae rebus falsis congruunt. Cum autem primis habuerint fidem, qualia sint ea, quae sequuntur, non circumspiciunt, sed defendunt omni modo; cum debeant prima illa, utrumne vera sint, an falsa, ex consequentibus judicare.*
Quae igitur illos ad Antipodas ratio perduxit? Videbant siderum cursus in occasum meantium; solem atque lunam in eamdem partem semper occidere, atque oriri semper ab eadem. Cum autem non perspicerent, quae machinatio cursus eorum temperaret, nec quomodo ab occasu ad orientem remearent, coelum autem ipsum in omnes partes putarent esse devexum, quod sic videri, propter immensam latitudinem necesse est: existimaverunt, rotundum esse mundum sicut pilam, et ex motu siderum opinati sunt coelum volvi, sic astra solemque, cum occiderint, volubilitate ipsa mundi ad ortum referri. Itaque et aereos orbes fabricati sunt, quasi ad figuram mundi, eosque caelarunt portentosis quibusdam simulacris, quae astra esse dicerent. Hanc igitur coeli rotunditatem illud sequebatur, ut terra in medio sinu ejus esset inclusa. Quod si ita esset, etiam ipsam terram globo similem; neque enim fieri posset, ut non esset rotundum, quod rotundo conclusum teneretur. Si autem rotunda etiam terra esset, necesse esse, ut in omnes coeli partes eamdem faciem gerat, id est montes erigat, campos tendat, maria consternat Quod si esset, etiam sequebatur illud extremum, ut nulla sit pars terrae, quae non ab hominibus caeterisque animalibus incolatur. Sic pendulos istos Antipodas coeli rotunditas adinvenit. Quod si quaeras ab iis, qui haec portenta defendunt, quomodo non cadunt omnia in inferiorem illam coeli partem; respondent, hanc rerum esse naturam, ut pondera in medium ferantur, et ad medium connexa sint omnia, sicut radios videmus in rota; quae autem levia sunt, ut nebula, fumus, ignis, a medio deferantur, ut coelum petant. Quid dicam de iis nescio, qui, cum semel aberraverint, constanter in stultitia perseverant, et vanis vana defendunt; nisi quod eos interdum puto, aut joci causa philosophari, aut prudentes et scios mendacia defendenda suscipere, quasi ut ingenia sua in malis rebus exerceant, vel ostendant. At ego multis argumentis probare possam, nullo modo fieri posse, ut coelum terra sit inferius, nisi et liber jam concludendus esset, et adhuc aliqua restarent, quae magis sunt praesenti operi necessaria. Et quoniam singulorum errores percurrere non est unius libri opus, satis sit pauca enumerasse, ex quibus possit qualia sint caetera intelligi (Eng. trans. by John J. Savage, from *The Divine Institutes, Books I–VII*, The Fathers of the Church, vol. 49 (Washington, DC: Catholic University of America Press), pp. 229-230).

Fathers of the Church with regard to the sphericity of the Earth and the Antipodes.[17] The subject of the waters above the firmament is not brought into question by him. One has the impression that the narration of the Bible *must* be accepted in its literal version.

From what we know from the history of the subsequent centuries, Lactantius's work has always enjoyed great respect and consideration in the world of the institutions of the Church, so much so that scholars of the past often put it together with the work of St. Augustine, although modern scholars do not do so. As a confirmation of this assertion, we can quote the fact that one of the first works printed in Italy (at Subiaco, in the Benedictine monastery of St. Scolastica) by the German monks-typographers Schweynheym and Pannartz in 1465 was precisely the *Divinae institutiones*.

Even Copernicus, in the Preface to his work *De Revolutionibus orbium coelestium* dedicated to the Pope Paul III, having to quote an author who, ignorant of the mathematical sciences, "speaks quite childishly about the earth's shape", quotes Lactantius: "Perhaps there will be babblers who claim to be judges of astronomy although completely ignorant of the subject and, badly distorting some passage of Scripture to their purpose, will dare to find fault with my undertaking and censure it. I disregard them even to the extent of despising their criticism as unfounded. For it is not unknown that Lactantius, otherwise an illustrious writer but hardly an astronomer, speaks quite childishly about the earth's shape, when he mocks those who declared that the earth has the form of a globe. Hence scholars need not be surprised if any such persons will likewise ridicule me. Astronomy is written for astronomers".[18]

Even Voltaire, in his *Dictionnaire Philosophique*, at the entry "*Le ciel des anciens*",[19] unites Lactantius with St. Augustine when expressing a satirical remark about the ancients' thought on the Earth's shape.

Perhaps because the Latin prose of Lactantius was very elegant, with a Ciceronian style (in fact he was nicknamed the Christian Cicero), his work has enjoyed a very lasting reputation and respect, even if not comparable with its doctrinal prominence.

1.4 Basil the Great (329–379)

In the fourth century, in the East there were three prominent Fathers of the Church (sometimes called "the three luminaries of Cappadocia"): Basil of Caesarea (329–379), the friend of his Gregory of Nazianzus (330–390) and his brother Gregory of Nyssa (335–394). We shall only deal with the first one, with the only and obvious aim of talking about his homilies on Genesis. Let us begin by providing some brief biographical information. Basil was born at Caesarea in Cappadocia in 329 into a Christian family (his father was a rich rhetorician and lawyer) and received a Christian education on the part of his grandmother Macrina at

---

[17] William H. Stahl, *Roman Science* (The University of Wisconsin Press, 1962).

[18] Nicolai Copernici Torinensis [Nicholas Copernicus], *De Revolutionibus orbium coelestium, libri VI* (Nurimberg: apud Joh. Petreius, 1543) (Eng. trans. by Edward Rosen, from Nicolaus Copernicus, *On the Revolutions* (Baltimore: Johns Hopkins University Press, 1992), p. 5).

[19] Let us report a brief excerpt of Voltaire's text: "St. Austin calls the notion of antipodes an absurdity; and Lactantius flatly says, 'Are there any so foolish as to believe there are men whose head is lower than their feet?' St. Chrysostom, in his fourteenth homily, calls out, 'Where are they who say the heavens are moveable, and their form round?' Lactantius again says, b. iii, of his *Institutions*, 'I could prove to you by a multitude of arguments, that it is impossible the heavens should encompass the earth'. The author of *Spectacle de la Nature* is welcome to tell the chevalier over and over, that Lactantius and Chrysostom were eminent philosophers; still it will be answered that they were great saints, which they may be without any acquaintance wilh astronomy. We believe them to be in heaven, but own that in what part of the heavens they are we know not" (Eng. trans. from Voltaire, *The philosophical Dictionary* (London: Wynne and Scholey, 1802), pp. 190-191).

Neocaesarea on Pontus. Afterwards he studied at Constantinople and then at Athens, where he befriended Gregory of Nazianzus. Around 360, Eusebius, bishop of Caesarea, ordained him and in 370, after Eusebius's death, he was elected bishop.

In his activity as a theologian, he had to fight Arianism for a long time, since it was the dominant heresy in the fourth century. Like other Fathers of the Church, Basil was also engaged in an intense activity of preaching. Several homilies by him are extant (nine of them concerning the *Hexaemeron* and twenty-four on different subjects).[20] As we shall see, in the nine homilies on the six days of the creation of the world, he uses the scientific doctrines of Antiquity, especially those of Aristotle, many times even quarrelling bitterly with them.

Like Origen before him, Basil too had an audience of common people and this conditioned both the language and the times of the homilies. He says in the third homily of the *Hexaemeron*: "I know that many artisans, belonging to mechanical trades, are crowding around me. A day's labour hardly suffices to maintain them; therefore I am compelled to abridge my discourse, so as not to keep them too long from their work. What shall I say to them? The time which you lend to God is not lost: he will return it to you with large interest".[21]

Over the centuries, scholars have much insisted on the fact that Basil, in his exegeses of Genesis, was inclined to a literal interpretation, entering into debate with the exegetes who were inclined to an allegorical interpretation (that is, Origen). In that regard and in support of the interpretation *ad litteram* (as we shall see, St. Augustine will bring up the same question again) two excerpts are usually quoted, one from the first homily and the other from the ninth. We bring forward both of them, as a preliminary assay of Basil's thought: "'Darkness was upon the face of the deep.' A new source for fables and most impious imaginations if one distorts the sense of these words at the will of one's fancies. By 'darkness' these wicked men do not understand what is meant in reality—air not illumined, the shadow produced by the interposition of a body, or finally a place for some reason deprived of light. For them 'darkness' is an evil power, or rather the personification of evil, having his origin in himself in opposition to, and in perpetual struggle with, the goodness of God".[22] At last, he says in the ninth homily: "I know the laws of allegory, though less by myself than from the works of others. There are those truly, who do not admit the common sense of the Scriptures, for whom water is not water, but some other nature, who see in a plant, in a fish, what their fancy wishes, who change the nature of reptiles and of wild beasts to suit their allegories, like the interpreters of dreams who explain visions in sleep to make them serve their own ends. For me grass is grass; plant, fish, wild beast, domestic animal, I take all in the literal sense. 'For I am not ashamed of the gospel.' Those who have written about the nature of the universe have discussed at length the shape of the earth. If it be spherical or cylindrical, if it resemble a disc and is equally rounded in all parts, or if it has the forth of a winnowing basket and is hollow in the middle; all these conjectures have been suggested by cosmographers, each one upsetting that of his predecessor. It will not lead me to give less importance to the creation of the universe, that the servant of God, Moses, is silent as to shapes; he has not said that the earth is a hundred and eighty thousand furlongs in circumference; he has not measured into what extent of air its shadow projects itself whilst the sun revolves around it, nor stated how this shadow, casting itself upon the moon, produces eclipses. He has passed over in silence, as useless, all that is unimportant for us. Shall I then

---

[20] See, for instance, Basilio di Cesarea, *Omelie sull'esamerone e di argomento vario*, ed. Francesco Trisoglio (Bompiani, 2017).

[21] Philip Schaff and Henry Wallace, eds. *Nicene and Post-Nicene Fathers Second Series: Volume VIII Basil* (New York: Cosimo Classics, 2007), Homily III.1, p. 65.

[22] Schaff and Wallace, *Nicene and Post-Nicene Fathers*, op. cit., Homily II.4, pp. 60-61.

prefer foolish wisdom to the oracles of the Holy Spirit? Shall I not rather exalt Him who, not wishing to fill our minds with these vanities, has regulated all the economy of Scripture in view of the edification and the making perfect of our souls? It is this which those seem to me not to have understood, who, giving themselves up to the distorted meaning of allegory, have undertaken to give a majesty of their own invention to Scripture. It is to believe themselves wiser than the Holy Spirit, and to bring forth their own ideas under a pretext of exegesis. Let us hear Scripture as it has been written".[23]

As it can be seen from the last quoted excerpt, Basil shows a knowledge of the results obtained by the Greek philosophers with regard to the Earth. Let us see now how this knowledge is used in the interpretation of the verses on the waters above the firmament. Let us start from III.7: "Therefore we read: 'Let there be a firmament in the midst of the waters, and let it divide the waters from the waters.' I have said what the word firmament in Scripture means. It is not in reality a firm and solid substance which has weight and resistance; this name would otherwise have better suited the earth. But, as the substance of superincumbent bodies is light, without consistency, and cannot be grasped by any one of our senses, it is in comparison with these pure and imperceptible substances that the firmament has received its name. Imagine a place fit to divide the moisture, sending it, if pure and filtered, into higher regions, and making it fall, if it is dense and earthy; to the end that by the gradual withdrawal of the moist particles the same temperature may be preserved from the beginning to the end. You do not believe in this prodigious quantity of water; but you do not take into account the prodigious quantity of heat, less considerable no doubt in bulk, but exceedingly powerful nevertheless, if you consider it as destructive of moisture. It attracts surrounding moisture, as the melon shows us, and consumes it as quickly when attracted, as the flame of the lamp draws to it the fuel supplied by the wick and burns it up. Who doubts that the aether is an ardent fire? If an impassable limit had not been assigned to it by the Creator, what would prevent it from setting on fire and consuming all that is near it, and absorbing all the moisture from existing things? The aerial waters which veil the heavens with vapours that are sent forth by rivers, fountains, marshes, lakes, and seas, prevent the aether from invading and burning up the universe. Thus we see even this sun, in the summer season, dry up in a moment a damp and marshy country, and make it perfectly arid. What has become of all the water? Let these masters of omniscience tell us. Is it not plain to everyone that it has risen in vapour, and has been consumed by the heat of the sun? They say, none the less, that even the sun is without heat. What time they lose in words! And see what proof they lean upon to resist what is perfectly plain. Its colour is white, and neither reddish nor yellow. It is not then fiery by nature, and its heat results, they say, from the velocity of its rotation. What do they gain? That the sun does not seem to absorb moisture? I do not, however, reject this statement, although it is false, because it helps my argument. I said that the consumption of heat required this prodigious quantity of water. That the sun owes its heat to its nature, or that heat results from its action, makes no difference, provided that it produces the same effects upon the same matter. If you kindle fire by rubbing two pieces of wood together, or if you light them by holding them to a flame, you will have absolutely the same effect. Besides, we see that the great wisdom of Him who governs all, makes the sun travel from one region to another, for fear that, if it remained always in the same place, its excessive heat would destroy the order of the universe. Now it passes into southern regions about the time of the winter solstice, now it returns to the sign of the equinox; from thence it betakes itself to northern regions during the summer

----

[23] Schaff and Wallace, *Nicene and Post-Nicene Fathers*, op. cit., Homily IX.1, pp. 101-102. In this regard, we also refer the reader to Mario Girardi, *Basilio di Cesarea Interprete della Scrittura. Lessico, principi ermeneutici* (Bari: EdiPuglia, 1988).

solstice, and keeps up by this imperceptible passage a pleasant temperature throughout all the world.

"Let the learned people see if they do not disagree among themselves. The water which the sun consumes is, they say, what prevents the sea from rising and flooding the rivers; the warmth of the sun leaves behind the salts and the bitterness of the waters, and absorbs from them the pure and drinkable particles, thanks to the singular virtue of this planet in attracting all that is light and in allowing to fall, like mud and sediment, all which is thick and earthy. From thence come the bitterness, the salt taste and the power of withering and drying up which are characteristic of the sea. While as is notorious, they hold these views, they shift their ground and say that moisture cannot be lessened by the sun".[24]

Rather than saying what the firmament is, Basil essentially says what it is not and then passes to argue with those (even if he does not quote him, his target is certainly Aristotle's *Meteorologica*, I,341a17) who maintain that the stars, including the sun, are not warm by their nature. In fact Aristotle maintains that the heat is an affection of the sensation and not owing to the sun.

Basil seems instead to base himself on Posidonius (135 BC-51 BC) who maintains that "the sun is fire in its pure state … it is fire since it produces all that, by its nature, the fire produces".[25]

Finally, with regard to the separation of the waters, in III.9, Basil writes: "But as far as concerns the separation of the waters I am obliged to contest the opinion of certain writers in the Church who, under the shadow of high and sublime conceptions, have launched out into metaphor, and have only seen in the waters a figure to denote spiritual and incorporeal powers. In the higher regions, above the firmament, dwell the better; in the lower regions, earth and matter are the dwelling place of the malignant. So, say they, God is praised by the waters that are above the heaven, that is to say, by the good powers, the purity of whose soul makes them worthy to sing the praises of God. And the waters which are under the heaven represent the wicked spirits, who from their natural height have fallen into the abyss of evil. Turbulent, seditious, agitated by the tumultuous waves of passion, they have received the name of sea, because of the instability and the inconstancy of their movements. Let us reject these theories as dreams and old women's tales. Let us understand that by water water is meant; fot the dividing of the waters by the firmament let us accept the reason which has been given us. Although, however, waters above the heaven are invited to give glory to the Lord of the Universe, do not let us think of them as intelligent beings; the heavens are not alive because they 'declare the glory of God', nor the firmament a sensible being because it 'sheweth His handiwork'. And if they tell you that the heavens mean contemplative powers, and the firmament active powers which produce good, we admire the theory as ingenious without being able to acknowledge the truth of it. For thus dew, the frost, cold and heat, which in Daniel are ordered to praise the Creator of all things, will be intelligent and invisible natures. But this is only a figure, accepted as such by enlightened minds, to complete the glory of the Creator. Besides, the waters above the heavens, these waters privileged by the virtue which they possess in themselves, are not the only waters to celebrate the praises of God. 'Praise the Lord from the earth, ye dragons and all deeps.' Thus the singer of the Psalms does not reject the deeps which our inventors of allegories rank in the divisions of evil; he admits them to the universal choir of creation, and the deeps sing in their language a harmonious hymn to the glory of the Creator".[26]

---

[24] Schaff and Wallace, *Nicene and Post-Nicene Fathers*, op. cit., Homily III.8, pp. 69-70.

[25] See Diogenes Laertius, *Lives of Eminent Philosophers* VII.144.

[26] Schaff and Wallace, *Nicene and Post-Nicene Fathers*, op. cit., Homily III.9, pp. 71.

Above (III, 4), he had said: "Before laying hold of the meaning of Scripture let us try to meet objections from other quarters. We are asked how, if the firmament is a spherical body, as it appears to the eye, its convex circumference can contain the water which flows and circulates in higher regions? What shall we answer? One thing only: because the interior of a body presents a perfect concavity it does not necessarily follow that its exterior surface is spherical and smoothly rounded. Look at the stone vaults of baths, and the structure of buildings of cave form; the dome, which forms the interior, does not prevent the roof from having ordinarily a flat surface. Let these unfortunate men cease, then, from tormenting us and themselves about the impossibility of our retaining water in the higher regions".[27]

For the moment, one could draw from this the conclusion that Basil doesn't feel up to denying the spherical shape of the Earth (as Lactantius did), together with the Aristotelian cosmography, and then he is obliged to conceptual acrobatics for reconciling the Holy Scripture and the science of that time.

If we wanted to express a consideration, even if temporary, about what we have seen until now, we could say that within the fourth century all possible currents of thought aimed at the interpretation of the first verses of Genesis saw the light. Wanting to simplify, substantially three are the ways of interpreting the famous verses. The most ancient, which, as we have seen, dates back to Philo of Alexandria and has enjoyed so much fortune up to the present time, is given by the allegorical interpretation. Alternatively, we have the interpretation (which is not really an interpretation, but an acceptance of the Bible's text in its literal meaning) after the fashion of Lactantius. We want to say that it is not only the "letter" of the cosmogonic narration of Genesis to be accepted, but also the cosmography on which it is centered, that is, the Earth is imagined as a flat disc. Finally, the attempt, more challenging and bristling with difficulties, of reconciling the Bible's text with the results of Greek science (corroborated in the subsequent centuries), which had established the sphericity of the Earth. Basil's work can certainly be considered the archetype of this type of interpretation, in the same way (always remaining in the field of the Christian exegesis) as that of Origen can be considered the archetype of the allegorical interpretation.

From now on we shall essentially deal with the developments that the question has undergone in the West, that is, in the treatments in the Latin language, since Latin from a certain time on became the language of the Church and, throughout the the Middle Ages, the question was discussed within the institutions of the Church.

1.5 St. Ambrose (339 ca.–397)

Around the midst of the fourth century, we find in the West a Father of the Church destined to leave his mark on the Church itself and its development. He is Aurelius Ambrosius, who has gone down in history as St. Ambrose. The importance of St. Ambrose in the history of the Church is not so much due to the profundity of his philosophical and theological doctrine as to his action of organizing the liturgy and his pastoral activity, through both his works and his writings.

We are interested in his activity of exegesis of the Bible and of Genesis in particular, since, as Tullio Gregory says, "With Ambrose and Augustine—but with a constant reminiscence of Origen—the hermeneutic lines of all the exegeses of the Dark Ages regarding the supracelestial waters are established: where the literal and allegorical interpretations alternate and overlap

---

[27] Schaff and Wallace, *Nicene and Post-Nicene Fathers*, op. cit., Homily III.4, pp. 67.

each other, now reopening the various hypotheses on the particular conditions of the existence of the waters beyond the celestial vault, now following the paths of the allegorical and typological interpretation with different issues according with the infinite possibilities provided by the meanings of the Bible; here prevailing is the identification of the supracelestial waters with the angels, founded on the verses of the Psalms and of Daniel to which one has often referred; rarer is the identification of the waters under the firmament with the powers of the evil and the demons, even if the opposition was instead present between evil and good, saints and sinners".[28]

As usual, in order to contextualize Ambrose in the Roman world of his time (during his life nine emperors followed each other, from Constant II to Theodoric I), we supply some brief biographical information. Ambrose was born in Treves in 339 or 340 into an important Roman senatorial family of the Catholic religion (his father held an important position in Treves). On his father's premature death he followed his mother and went to Rome, where he attended the best schools. Afterwards he was initiated into an administrative career and in 370 he was sent in Milan in the office of governor. He met with success in his work and became an outstanding figure at the court of the emperor Valentinian. In 374, on the death of the bishop Auxentius (an ardent Arian), he found himself forced to pacify the contrast between the Arians and the non-Arians. His peacemaking gesture was so effective that he was elected as bishop by popular acclaim. In this case, the principal of apostolic succession was not applied! Even though at the beginning he was reluctant to accept the position (irregular enough!), he was obliged to yield to the pressure of the emperor, who was in his turn pressured by the people. At that point, Ambrose convinced himself that that it must be the will of the Lord. In this way, within a week, he was baptized and, on 7 December 374 he was ordained bishop.

The culture he had acquired when he was young (he had also studied the Greek) helped him in studying the Bible and theology in depth. Among the writings of Ambrose, particularly important are *De Officiis ministrorum* (which takes inspiration from Cicero's *De Officiis*) and the exegetic works *Expositio evangelii secundum Lucam* (a commentary on the gospel of Luke) and *Hexaemeron*[29] (which collects nine homilies in six books addressed to the six days of the creation. We shall deal with this last work, particularly with the third homily. In fact, it is precisely in the third homily that the verses regarding the second day of the creation are commented upon. The scholars of the *Hexaemeron* have pointed out how much Ambrose's work owes to that of Basil, even in certain little details. Indeed, comparing the two works, one often has the impression that many among the Ambrose's talks are a paraphrase of those of Basil, but with the presence of a new element. In fact, Ambrose belonged to the elite of the Roman world of his time, in particular to the educated elite, therefore his talks are reminiscent of the works which we nowadays call "the Latin classics". In the commentary on the first day (that is, in the first two homilies) we find references (we call them in this way, since the authors

---

[28] *"Con Ambrogio e Agostino—ma anche con un constante ricordo di Origene—sono fissate le linee ermeneutiche di tutta l'esegesi altomedievale relativa alle acque sopra-celesti: ove si alternano e si sovrappongono le interpretazioni letterali a quelle allegoriche, riproponendo ora le varie ipotesi sulle particolari condizioni dell'esistenza delle acque al di là della volta celeste, ora seguendo i sentieri dell'interpretazione allegorica e tipologica con esiti diversi secondo le infinite possibilità offerte dai sensi della Bibbia; qui prevalente è l'identificazione delle acque sopracelesti con gli angeli, fondata sui versetti dei Salmi e di Daniele cui si è spesso fatto riferimento; più rara è l'identificazione delle acque sotto il cielo con le potenze del male e i demoni, pur presentandosi invece l'opposizione bene e male, santi e peccatori".* From: Tullio Gregory, "Le acque sopra il firmamento. Genesi e tradizione esegetica", in: *L'acqua nei secoli altomedievali*, vol. I (Spoleto: Fondazione Centro italiano di studi sull'alto Medioevo, 2008), p. 19 (our Eng. trans.).

[29] See Saint Ambrose, *Hexaemeron, Paradise, and Cain and Abel*, trans. John J. Savage. Fathers of the Church: A New Translations, vol. 42 (Washington DC: Fathers of the Church, 1961).

are not explicitly quoted) to Cicero (*De Natura Deorum, De Senectute*), Virgil (*Georgics, Aeneid*), Lucretius (*De rerum natura*). Therefore, the culture of the milieu of origin differentiates the bishop of Milan from Basil, even if they are unified by the deep knowledge of the Bible. A thing which we should still further point out is that the knowledge of the theories of the Greek philosophers, which Ambrose shows, appears to be "second hand", that is, derived from the Roman popularizations, even if to call the *Somnium Scipionis* (in the sixth book of Cicero's *Res publica*) a popularization of Posidonius's cosmology may appear disrespectful for that magnificent Latin prose.

Let us see now how Ambrose, in actual fact following Basil's example, replies to the interpreters with whom he argues about the fundamental question: "And first of all these interpreters wish to destroy the profound impressions which frequent reading of the Scriptures have made in our mind, maintaining that waters cannot exist above the heavens. That heavenly sphere, they say, is round, with the earth in the middle of it; hence, water cannot stay on that circular surface, from which it needs must flow easily away, falling from a higher to a lower position. For how, they say, can water remain on a sphere when the sphere itself revolves?

"This is one of those sophistical arguments of the subtlest kind. Grant me an opportunity to reply. If it is not granted, there need be no further room for discussion.

"They ask us to concede to them that heaven turns on its axis with a swift motion, while the sphere of the earth remains motionless, so as to conclude that waters cannot stay above the heavens, because the axis of heaven as it revolved would cause these to flow off. They wish, in fact, that we grant them their premise and that our reply be based on their beliefs. In this way they would avoid the question of the existence of length and breadth in that height and depth, a fact which no one can comprehend except Him who is filled 'with the fullness of the Godhead,' as the Apostle says. For who can easily set himself up to be a judge of God's work? There exists, therefore, breadth in the very heights of heaven.

"To speak of matters within our knowledge, there are a great many buildings which are round in the exterior but are square-shaped within, and vice-versa. These buildings have level places on top, where water usually collects. We are led to mention these matters in order to draw the attention of these interpreters to the fact that their opinions can be confuted by other opinions closer to the truth and that they may cease measuring such a mighty work of God in terms of human work and merely on an estimate of our own capacities".[30]

And, as regards what we can see, Ambrose does not field the question with valid reasons. And he continues: "We follow the tradition of the Scriptures and we value the work by our esteem of the Author, as to what was said, who said it, and to whom it was said. 'Let there be a

---

[30] St. Ambrose, *Hexaemeron* III, 9: *Et primo uolunt id destruere quod frequenti scripturarum lectione inolitum nostris et inpressum est mentibus, quia aquae super caelos esse non possunt, dicentes rotundum esse orbem illum caeli, cuius in medio terra sit, et in illo circuitu aquam stare non posse, quod necesse est defluat et labatur, cum de superioribus ad inferiora decursus est. Quomodo enim aqua super orbem stare ut aiunt potest, cum orbis ipse uoluatur? Haec est illa uersutia dialecticae. Da mihi unde tibi.respondeam. Quod si non detur, nullum uerbum refertur. Petunt sibi concedi axem caeli torqueri motu concito, orbem autem terrae esse inmobilem, ut astruant aquas super caelos esse non posse, quod omnes eas uoluendo se axis effunderet, quasi uero, ut concedamus illis quod postulant et secundum eorum opiniones illis respondeam, negare possint in illa altitudine et profundo uel longitudinem esse et latitudinem, quam nemo potest conprehendere nisi is qui inpletur in omnem plenitudinem dei, ut apostolus ait. Quis enim facile poterit esse diuini operis aestimator? Est ergo latitudo in ipsa caeli altitudine. Sunt etiam, ut de bis dicamus quae scire possumus, pleraque aedificia foris rotunda, intus quadrata et foris quadrata, intus rotunda, quibus supcriora plana sunt, in quibus aqua haerere soleat. Quae tamen ideo dicimus, ut aduertant opiniones suas opinionibus ueri similioribus reuinci posse et desinant tantum opus dei humanae operationis et nostrae possibilitatis contemplatione metiri* (Eng. trans. John J. Savage, op. cit., pp. 52-53).

firmament made,' He said, 'amidst the waters and let it divide the waters from the waters.' From this I learn that the firmament is made by a command by which the water was to be separated and the water above be divided from the water below. What is clearer than this?

"He who commanded the waters to be separated by the interposition of the firmament lying between them provided also the manner of their remaining in position, once they were divided and separated. The word of God gives nature its power and an enduring quality to its matter, as long as He who established it wishes it to be so, as it is written: 'He hath established them forever and for ages of ages. He hath made a decree and it shall not pass away'".[31] And finally: "But why is it impossible for Him who gave strength to the weak, so that they could say: 'I can do all things in him who strengthens me.'

"Let them tell us whether, when 'the air thickens into cloud' rain is then produced by clouds or whether it is collected in the lap of the clouds? We so frequently see clouds issuing from the mountains, I ask you: Does the water rise from the earth or does the water which is over the heavens fall in copious rain? If water rises, it surely is against nature that the element which is heavier should be borne to a higher place and that it should be carried there by air, although this is a lighter element. Or if water is whirled by the rapid motion of the entire world system, in that case it is absorbed from the lowest sphere and, likewise, it is poured forth from the highest. If it does not cease to be poured forth, as they claim, surely it does not cease to be absorbed, because, if the axis of the heavens is ever in movement, the water, too, is always being absorbed. If water descends, then it is clear that it is continuously above the heavens in a position from which it can flow downwards.

"What prevents us, then, from admitting that water is suspended above the heavens? How can they say that the earth, although it is certainly heavier than water, stays suspended and immobile in the middle? Following the same principle, they can admit the water which is above the heavens does not descend because of the rotation of that celestial sphere. Just as the earth is suspended in the void and stays immobile in position, its weight being balanced on every side, in like manner the water, too, is balanced by weights either equal to or greater than that of the earth. For the same reason, the sea does not tend to inundate the land without a special command to do so."[32]

As a conclusion, he returns to the nature of the firmament: "But let us return to our theme: 'Let there be a firmament made amidst the water.' Let it not disturb you, as I have already said,

---

[31] Ibidem, III, 10: *Nos autem scripturarum seriem atque ordinem sequimur et opus contemplatione aestimamus auctoris, quid dictum sit et quis dixerit et cui dixerit.* Fiat *inquit jirmamentum in medio aquae et sit discernens inter aquas. Audio firmamentum fieri praecepto, quo diuideretur aqua et ab inferiore superior discerneretur. Quid hoc manifestius? Qui iussit discerni aquam interiecto et medio firmamento prouidit quemadmodum diuisa atque discreta manere possit. Sermo dei uirtus naturae est et diuturnitas substantiae, quoad uelit eam manere qui statuit, sicut scriptum est:* Statuit ea in saeculum et in saeculum saeculi; praeceptum posuit, et non praeteribit (Eng. trans. John J. Savage, op. cit., p. 53).

[32] Ibidem, III, 11: *...sed quid* impossibile *ei qui dedit posse infirmis, ut infirmus dicat:* Omnia possum in eo qui me confortat? *Dicant certe quemadmodum aer cogatur in nubem, utrum pluuia nubibus generetur an sinu nubium colligatur. Videmus plerumque exire nubes de montibus. Quaero utrum de terris ascendat aqua an ea quae super caelos est largo imbre descendat. Si ascendit, utique contra naturam est, ut ascendat in superiora quae grauior est et portetur aere, cum aer subtilior sit. Aut si conciti orbis totius motu rapitur aqua, sicut imo orbe rapitur ita summo orbe diffunditur. Si fundi, ut uolunt, non desinit, utique non desinit rapi, quia si axis caeli semper mouetur, et aqua semper hauritur. Si descendit, manet ergo iugiter supra caelos, quae habet unde descendat. Deinde quid obstat, si confiteantur quia aqua super caelos suspensu sit? Num quo uerbo dicunt terram in medio esse suspensam et immobilem manere, cum utique grauior sit quam aqua? Ea ratione possunt dicere non praecipitari aquam orbis illius caelestis conuersione, quae super caelos est. Sicut enim terra <in> inani suspenditur uel pendere librato undique immobilis perseuerat, ita et aqua aut grauioribus aut aequis cum terra ponderibus examinatur. Ideoque non facile superfunditur mare terris, nisi cum iubetur exire* (Eng. trans. John J. Savage, op. cit., pp. 55-56).

that above He speaks of heaven and here of a firmament, since David also says: 'The heavens narrate the glory of God and the firmament declareth the work of his hands.' That is to say, the created world, when one beholds it, praises its own Author, for His invisible majesty is recognized through the things that are visible. It seems to me that the word 'heaven' is a generic term, because Scripture testifies to the existence of very many heavens. The word 'firmament' is more specific, since here also we read: 'And he called the firmament, heaven'. In a general way, He would seem to have said above that heaven was made in the beginning so as to take in the entire fabric of celestial creation, and that here the specific solidity of this exterior firmament is meant. This is called the firmament of heaven, as we read in the prophetic hymn, 'Blessed are thou in the firmament of heaven.' For heaven, which in Greek is called οὐρανός, in Latin, *caelum,* is connected with the word 'stamped' (*caelatum*),because the heavens have the lights of the stars impressed on them like embossed work, just as silver plate is said to be 'stamped' when it glitters with figures in relief. The word οὐρανός, seems to be derived from the Greek verb 'to be seen' [ὀρᾶσθαι]. In distinction, therefore, to the earth, which is darker, the sky is called οὐρανός, because it is bright, that is to say, visible. Hence, I believe, is the origin of that expression: 'The winged ones of heaven always behold the face of my Father, who is in heaven.' And again: 'The winged fowl above the firmament of heaven.' The powers which exist in that visible place behold all these things and have them subject to their observation".[33] Summarizing, we can say that two are the 'strong' arguments used by Ambrose: on the one hand, the power of God, who can even do the things one judges impossible (this "loophole" will often be used also in the future); on the other hand the quotations of other loci of the Bible where analogous things are maintained, that is, he explains the Bible with the Bible.

1.6 Augustine (354–430)

With Augustine (Aurelius Augustinus, 354–430) one can say that substantially the Patristics period concludes. While until the fourth century the theological elaboration of the Christian doctrine, at least in its main lines, also went on with the use of the Greek language, with Ambrose the use of the Latin language will become fundamental and Augustine will make the Latin the language of the Church.

The importance of the theological and philosophical work of Augustine (later St. Augustine), destined to influence the culture of the European Middle Ages, is indisputable and the bulk of his writings is enormous. As in the other cases, we shall only deal with the question of the waters above the firmament, in which, and for a long time, Augustine himself was also involved. First of all, as usual, let us provide a brief biographical background.

Aurelius Augustinus was born in 374 AD in Tagaste (Numidia, in Roman Africa, now

---

[33] Ibidem, III, 15: *Sed reuertamur ad propositum.* Fiat firmamentum inter medium aquae. *Non moueat, sicut iam dixi, quia supra caelum ait,* Hic dicit firmamentum, *quoniam et Dauid ait:* Caeli enarrant gloriam dei, et opera manuum eius adnuntiat firmamentum, *hoc est: mundi opus, cum uidetur, suum laudat auctorem; inuisibilis enim maiestas eius per ea quae uidentur agnoscitur. Et uidetur mihi nomen caelorum commune esse, quia plurimos caelos scriptura testificatur, nomen autem esse speciale firmamentum, siquidem et hic ita habet:* Et uocauit firmamentum caelum, *ut uideatur supra generaliter dixisse in principio caelum factum, ut omnem caelestis creaturae fabricam conprehenderet, hic autem specialem firmamenti huius exterioris soliditatem, quod dicitur caeli firmamentum, sicut legimus in hymno prophetico:* Benedictus es in firmamento caeli. *Nam caelum, quod* οὐρανός graece dicitur, *latine, quia inpressa stellarum lumina uelut signa habeat, tamquam caelatum appellatur, sicut argentum, quod signis eminentibus refulget, caelatum dicimus,* οὐρανός autem *ἀπὸ τοῦ* ὀρᾶσθαι *dicitur, quod uidetur.* Πρὸς ἀντιδιατολήν igitur terrae, quae obscurior est, οὐρανός nuncupatur, quia lucidus est, tamquam uisibilis. Vnde puto et illud dictum uolatilia caeli semper uident faciem patris mei, qui in caelis est et uolatilia circa firmamentum caeli, eo quod potestates, quae sunt in illo uisibili loco, spectent haec omnia et subiecta suis habeant conspectibus (Eng. trans. John J. Savage, op. cit., p. 53).

Algeria) His father Patrice was pagan and his mother Monica Christian, indeed very Christian. He carried out his university education with great success, first at Madaura and then at Carthage. He divided his early youth between his educational successes and the abuses of earthly pleasures (as he narrates in the *Confessions)*. It seems he was led to devote himself to the study of philosophy by reading a work by Cicero *(Hortensius,* now lost). He also resorted to the Bible (at that time available in a poor Latin translation, since St. Jeromes's translation would arrive later) and was fascinated by the Manichaean heresy, with its dualistic conception of the good and the evil, of which for nine years he was one of the more fretful followers. He began his teaching of rhetoric at first in Carthage, then in Rome and eventually in Milan. There, after having heard the sermons of Ambrose, he was definitively converted to Christianity. He gave up teaching and decided to go back to Africa and devote himself to a life of devoutness and study. His mother's death in Ostia, before his departure, kept him in Rome for one more year. By 388 he was already in Tagaste, where he lived with his friends for three years in a meditative idleness. In 392 he entered the priesthood and was ordained auxiliary bishop by Valerius. Upon Valerius's death (in 396) he was elected bishop of Hippo, where he remained until his death in 430.

A few years before his death, precisely from 426 to 427, Augustine wrote a singular work, which could be said unique in the history of literature, bearing the title *Retractationes* *(Retractions)*. In it the author expounds his critical remarks on his own works published until then; there were 93 works listed in order of publication.

Augustine says in the Prologue of the work, "For a long time I have been thinking about and planning to do something which I, with God's assistance, am now undertaking because I do not think it should be postponed: with a kind of judicial severity, I am reviewing my works—books, letters, and sermons—and, as it were, with the pen of a censor, I am indicating what dissatisfies me. For, truly, only an ignorant man will have the hardihood to criticize me for criticizing my own errors. But if he maintains that I should not have said those things which, indeed, dissatisfied me later, he speaks the truth and concurs with me. In fact, he and I are critics of the same thing, for I should not have criticized such things if it had been right to say them. But let each one, as he chooses, accept what I am doing".[34] He concludes, "For, perhaps, one who reads my works in the order in which they were written will find out how I progressed while writing. In order that this be possible, I shall take care, insofar as I can in this work, to acquaint him with this order".[35] Following Augustine's suggestion, we too shall make use of his work in order to see how the interpretation that he has given of Genesis 1:6-8 evolved over the course of more than thirty years.

The exegesis of Genesis was one of the principal interests of Augustine and in fact it appears in at least five works,[36] which we list in the order given by the author:

---

[34] St. Augustine, *Retractationum libri duo*, Prologus, M PL 34 col. 583: *Iam diu est ut facere cogito atque dispono quod nunc, adiuvante Domino, aggredior, quia differendum esse non arbitrar, ut opuscula mea, sive in libris sive in epistolis sive in tractatibus cum quadam iudiciaria severitate recenseam et, quod me offendit, velut censorio stilo denotem. Neque enim quisquam nisi impradens, ideo quia mea errata reprehendo, me reprehendere audebit. Sed si dicit non ea debuisse a me dici, quae postea mihi etiam displicerent, verum dicit et mecum facit. Eorum quippe reprehensor est, quorum et ego sum. Neque enim ea reprehendere deberem, si dicere debuissem. Sed, ut volet, quisque accipiat hoc quod facio* (Eng. trans. Mary Inez Bogan, in St. Augustine, *The Retractions*. Fathers of the Church: A New Translation, vol. 60 (New York: Fathers of the Church, 1968), p. 3).

[35] Ibidem, col. 586: *Quapropter quicumque ista lecturi sunt, non me imitentur errantem, sed in melius proficentem Inveniet enim fortasse quomodo scribendo profeterim, quisquis opuscula mea ordine quo scripta sunt legerit. Quod ut possit, hoc opere quantum potero curabo, ut eumdem ordinem noverit* (Eng. trans. Mary Inez Bogan, op. cit., p. 5).

[36] Cf. Gilles Palland, *Cinq études d'Augustin sur le début de la Genèse* (Bellarmin, 1972).

1) *De Genesi contra Manichaeos* (On Genesis, against the Manichees)
2) *De Genesi ad literam liber imperfectus* (On the literal interpretation of Genesis: An unfinished book)
3) *Confessiones* (Confessions)
4) *De Genesi ad litteram libri XII* (The literal meaning of Genesis in twelve books)
5) *De civitate Dei* (The city of God).

As we shall see, Augustine's thought, swinging between the allegorical interpretation and the literal (historic) one is also conditioned by the uncertainty regarding the conception of the physical world to which refer himself.

Let us begin with the first work in the list. This is the first of five treatises written by Augustine in the course of his fight against the Manichaeism. The subject we are interested in is dealt with in the eleventh chapter of the first book, where he says, "I do not recall that the Manichees are accustomed to find fault with this. The waters were divided so that some were above the firmament and others below the firmament. Since we said that matter was called water, I believe that the firmament of heaven separated the corporeal matter of visible things from the incorporeal matter of invisible things. For though heaven is a very beautiful body, every invisible creature surpasses even the beauty of heaven, and perhaps for that reason the invisible waters are said to be above the heaven. For few understand that they surpass the heaven, not by the places they occupy, but by the dignity of their nature, although we should not rashly affirm anything about this, for it is obscure and remote from the senses of men. Whatever the case may be, before we understand it, we should believe".[37]

Although Augustine maintains that he does not remember objections regarding verses 6-8 on the part of the Manichees, we can see that he feels bound nevertheless to speak his mind, which for the moment is limited to an allegorical interpretation. This chapter is not subject to retractations in the ninth chapter of the first book of the *Retractationes* regarding this work.

To introduce the later work, it is convenient to report what Augustine tells in the *Retractationes* (Book I, Chap. XVII): "After I had compared the two books of *On Genesis, against the Manichaeans,* and had explained the words of Scripture according to their allegorical meaning, not presuming to explain such great mysteries of natural things literally— that is, in what sense the statements there made can be interpreted according to their historical significance—I wanted to test my capabilities in this truly most taxing and difficult work also. But in explaining the Scriptures, my inexperience collapsed under the weight of so heavy a load and, before I had finished one book, I rested from this labor which I could not endure. But while I was re-examining my writings in the present work, this very book came into my hands, unfinished as it was, which I had not published and which I had decided to destroy since, at a later lime, I wrote twelve books entitled *On the Literal Meaning of Genesis.* Although in those books many questions seem to have been proposed rather than solved, yet this present book is by no means to be compared with those books. But, still, after I had re-examined this book, I decided to keep it so that it might serve as evidence, useful in my opinion, of my first attempts

---

[37] St. Augustine, *De Genesi contra Manichaeos*, I, XI, MPL 34: *Haec non memini Manichaeos reprehendere solere: tamen quod divisae sunt aquae ut aliae essent super firmamentum, et aliae sub firmamento, quoniam materiam illam dicebamus nomine aquae appellatam, credo firmamento coeli materiam corporalem rerum visibilium ab illa incorporali rerum invisibilium fuisse discretam. Cum enim coelum sit corpus pulcherrimum, omnis invisibilis creatura excedit etiam pulchritudinem coeli; et ideo fortasse super coelum esse dicuntur aquae invisibiles, quae a paucis intelleguntur non locorum sedibus, sed dignitate naturae superare coelum: quamquam de hac re nihil temere affirmandum est; obscura est enim, et remota a sensibus hominum: sed quoquo modo se habeat, antequam intellegatur, credenda est* (Eng. trans. Roland J. Teske in St. Augustine, *On Genesis: Two books on Genesis against the Manichees ; and, On the literal interpretation of Genesis, an unfinished book.* Fathers of the Church: A New Translation, vol. 84 (New York: Fathers of the Church, 1991), pp. 64-65).

to explain and search into the divine Scriptures, and I determined that its title should be *One Unfinished Book on the Literal Meaning of Genesis*".[38]

Here is the first attempt of interpretation *ad litteram*: "And God said, 'Let there be a firmament in the middle of the water, and let it divide the waters.' And so it was done. And God made the firmament and divided the water that was below the firmament from the water that was above the firmament. Were the waters above the firmament like these visible ones below the firmament? Scripture seems to refer to the water over which the Spirit was borne, and we took that water to be the matter of this world. Should we then believe that in this passage this matter is separated by the interposition of the firmament so that the lower matter is that of bodies and the higher matter that of souls? For Scripture here calls the firmament what it later calls heaven. Among bodies there is none better than the body of the heaven. Indeed heavenly bodies are completely different from earthly bodies, and the heavenly ones are better. I do not know how anything that surpasses their nature can still be called a body.

"Perhaps there is a power subject to reason, by reason of which God and the truth are known. This nature can be formed by virtue and prudence, and by their power its fluctuation is checked and bound. Hence, we can regard it as material, and it is rightly called water in Scripture, though it surpasses the reaches of the bodily heaven, not by stretches of space, but by the merit of its incorporeal nature. Since Scripture called heaven the firmament, we can without absurdity hold that anything below the ethereal heaven, in which everything is peaceful and stable, is more mutable and perishable and is a kind of corporeal matter prior to the reception of beauty and the distinction of forms; for this reason it is said to be below the firmament. There were some who believed that these visible and cold waters surrounded the surface of heaven. They tried to use as a proof [of this] the slowness of one of the seven wandering stars which is higher than the rest and is called *Phainōn* by the Greeks. It takes thirty years to complete its starry orbit. Its slowness is supposedly due to its proximity to the cold waters that are above the heaven. I do not know how this opinion can be defended by those who have searched out these matters most carefully. We should affirm none of these opinions rashly, but carefully and moderately discuss them all."[39]

---

[38] Augustine, *Retractationum libri duo*, I, XVII, M PL 32: *Cum de Genesi duos libros cantra Manichaeos condidissem, quoniam secundum allegoricam significationem Scripturae verba tractaveram, non ausus naturalium rerum tanta secreta ad litteram exponere, hoc est quemadmodum possent secundum historicam proprietatem quae ibi dicta sunt accipi, volui experiri in hoc quoque negotiosissimo ac difficillimo opere quid valerem; sed in Scripturis exponendis tirocinium meum sub tanta sarcinae mole succubuit, et nondum perfecto uno libro ab eo quem sustinere non poteram labore conquievi. Sed in hoc opere, cum mea opuscula retractarem, iste ipse ut erat imperfectus venit in manus, quem neque edideram et abolere decreveram, quoniam scripsi postea duodecim libros quorum titulus est: De Genesi ad litteram. In quibus quamvis multa quaesita potius quam inventa videantur, tamen eis iste nullo modo est comparandus. Verum et hunc posteaquam retractavi manere volui, ut esset index, quantum existimo, non inutilis rudimentorum meorum in enucleandis atque scrutandis divinis eloquiis; eiusque titulum esse volui: De Genesi ad litteram imperfectus* (Eng. trans. Mary Inez Bogan, op. cit., p. 5).

[39] Augustine, *De genesi ad litteram liber imperfectus*, VIII, 29, M P L 34: *Et dixit Deus*, Fiat firmamentum in medio aquae, et sit dividens inter aquam et aquam. Et sic factum est. Et fecit Deus firmamentum, et divisit inter aquam quae erat sub firmamento, et inter aquam quae erat supra firmamentum. *Utrum aquae tales sint supra firmamentum, quales sub firmamento istae visibiles? an quia illam aquam videtur significare, supra quam Spiritus ferebatur, et eam intellegebamus esse ipsam mundi materiam, haec etiam hoc loco firmamento interposito discreta credenda est, ut inferior sit materia corporalis, superior animalis? Hoc enim firmamentum dicit, quod coelum postea vocat. Coelesti autem corpore nihil est in corporibus alia melius. Alia quippe corpora coelestia, et terrestria; et utique coelestia meliora: quorum naturam quidquid transit, nescio quemadmodum corpus possit vocari; sed est fortasse vis quaedam subiecta rationi, qua ratione Deus veritasque cognoscitur: quae natura, quia formabilis est virtute atque prudentia, cuius vigore cohibetur eius fluitatio atque constringitur, et ob hoc quasi materialis apparet, recte aqua divinitus appellata est; non locorum spatio, sed merito naturae incorporeae coeli corporei ambitum excedens. Et quoniam coelum firmamentum vocavit, non absurde intellegitur quidquid infra*

He continues, "Scripture added, 'And God made the firmament and divided the water that was below the firmament from the water that was above the firmament'. Does this refer to the working on that matter in order that the body of the heaven might come to be? Or was it perhaps for the sake of variety that it did not say above what it said here so that the text of the narrative might not bore the reader? We need not scrupulously search out all these matters. Let each choose what he can; only let him not say something rashly and assert something as known when it is not. Let him recall that he is human and is investigating the works of God to the extent we are permitted".[40]

Taking the words from the writing of Augustine himself. we could say effectively that, also for the passage regarding the waters above the firmament, "*multa quaesita potius quamvis inventa videantur*", many questions seem to have proposed rather than solved.[41]

Moreover, in this case, we have to do with a commentary rather than with an interpretation. In the third of the works listed, *Confessions*, which was written between 397 and 400—that is, before he began the *De Genesi ad litteram libri XII*—Augustine comes back again to the problem of the exegesis of Genesis and, in particular, of the creation. As we have said, this problem kept his mind occupied throughout his whole life as a Christian and appears, overbearingly, in the last two books of the *Confessions*. There is not yet an interpretation *ad litteram* (even if Augustine's concept of interpretation *ad litteram* will never correspond with the meaning that usually is attributed to this expression). Let us see an example of it: "For this corporeal heaven is truly marvellous, this firmament between the water and the waters which thou didst make on the second day after the creation of light, saying, 'Let it be done,' and it was done. This firmament thou didst call heaven, that is, the heaven of this earth and sea which thou madest on the third day, giving a visible shape to the unformed matter which thou hadst made before all the days".[42]

Here enthusiasm for the act of creation prevails over all and there is no sign regarding the nature of the firmament and its possible solidity. It is in the first book of *De genesi ad litteram* that Augustine explains how a study on Genesis should be carried out by interpreting *ad litteram* the fundamental passages. This work (in twelve books) presents itself as an accurate explanatory commentary on the first three chapters of Genesis, from the beginning of creation to the expulsion of Adam and Eve from Paradise. The first five books cover the work of creation

---

aethereum coelum est, in quo pacata et firma sunt omnia, mutabilius esse et dissolubilius. Quod genus corporalis materiae ante acceptam speciem distinctionemque formarum, a qua sub firmamento nominata est, fuerunt qui crederent has visibiles aquas et frigidas superficiem coeli superamplecti. Et documentum adhibere conati sunt de tarditate stellae unius de septem vagantibus, quae superior est caeteris, et a Graecis φαινον *dicitur, et triginta annis peragit signiferum circulum, ut ob hoc tarda sit, quia est frigidis aquis vicinior, quae supra coelum sunt. Quae opimo nescio quemadmodum possit apud eos defendi, qui subtilissime ista quaesierunt. Nihii autem horum temere affirmandum, sed caute omnia modesteque sunt tractanda* (Eng. trans. Roland J. Teske, op. cit., pp. 165-166).

[40] Ibidem, IX, 30: *Cum autem additum est:* Et fecit Deus firmamentum, et divisit inter aquam quae erat sub firmamento, et aquam quae erat supra firmamentum; *ipsa operatio in illa materia, ut corpus coeli fieret, significata? An forte varietatis causa, ut textus sermonis in fastidium non veniret, supra non est positum quod intra positum est, et non oportet scrupulose omnia rimari? Eligat quis quod potest: tantum ne aliquid temere atque incognitum pro cognito asserat; memineritque se hominem de divinis operibus quantum permittitur quaerere* (Eng. trans. Roland J. Teske, op. cit., p. 167).

[41] See note 38 above.

[42] Augustine, *Confessiones, liber duodecimus* 8.8: *Valde enim mirabile hoc caelum corporeum, quod firmamentum inter aquam et aquam secundo die post conditionem lucis dixisti: fiat, et sic est factum. Quod firmamentum vocasti caelum, sed caelum terrae huius et maris, quae fecisti tertio die dando speciem visibilem informi materiae, quam fecisti ante omnem diem* (Eng. trans. by Albert C. Outler, in St. Augustine, *Confessions and the Enchiridion* (Philadelphia: Westminster Press, 1955), p. 275).

and God's rest of the seventh day. For the subject we are interested in, we can limit ourselves to the first two books.

First of all we recall that, according to scholars, Augustine was certainly familiar with the *Homilies on Genesis* by Origen and with Basil's *Homilies in Hexaemeron*, in the Latin translation by Rufinus the former and by Eustathius the latter, since Augustine at that time was not familiar with the Greek (afterwards, in maturity, he studied it). He also knew the works of Tertullian, Lactantius and Ambrose. This to say that he had at his disposal antecedents for a comparison. Maybe this is the reason why he devotes so much space to explain the principles to be adopted in the exegesis of Genesis and in the discussion of the fundamental words which appear therein (beginning, heaven, earth, water, unformed matter, light, day and night). In fact, at the beginning of the first book, he asks himself if both types of interpretation (*ad litteram* and allegoric) can be admissible and opts for a positive answer resting on St. Paul: "In all the sacred books, we should consider the eternal truths that are taught, the facts that are narrated, the future events that are predicted, and the precepts or counsels that are given. In the case of a narrative of events, the question arises as to whether everything must be taken according to the figurative sense only, or whether it must be expounded and defended also as a faithful record of what happened. No Christian will dare say that the narrative must not be taken in a figurative sense. For St. Paul says: *Now all these things that happened to them were symbolic.* And he explains the statement in Genesis, *And they shall be two in one flesh*, as a great mystery in reference to Christ and to the Church".[43]

As an example, speaking about water, he says, "We might say that by the term 'water' the sacred writer wished to designate the whole of material creation. In this way he would show whence all things that we can recognize in their proper kinds had been made and formed, calling them water, because we observe all things on earth being formed and growing into their various species from moisture. Or we might say that by this term he wished to designate a certain kind of spiritual life, in a fluid state, so to speak, before receiving the form of its conversion".[44]

Then he resorts to the allegorical interpretation of the expression "And evening was made and morning made, one day". He rejects the interpretation of the evening as signifying the sin of the rational nature and of the morning as its renewal. In fact, he says, "But this is to give an allegorical and prophetical interpretation, a thing which I did not set out to do in this treatise. I have started here to discuss Sacred Scripture according to the plain meaning of the historical facts, not according to future events which they foreshadow. How, then, in the account of the creation and formation of things can we find evening and morning in the created spiritual light?

---

[43] Augustine, *De Genesi ad litteram libri duodecim* I, 1, 1: *In Libris autem omnibus sanctis intueri oportet quae ibi aeterna intimentur, quae facta narrentur, quae futura praenuntientur, quae agenda praecipiantur vel admoneantur. In narratione ergo rerum factarum quaeritur utrum omnia secundum figurarum tantummodo intellectum accipiantur, an etiam secundum fidem rerum gestarum asserenda et defendenda sint. Nam non esse accipienda figuraliter, nullus christianus dicere audebit, attendens Apostolum dicentem:* Omnia autem haec in figura contingebant illis *: et illud quod in Genesi scriptum est:* Et erunt duo in carne una, *magnum sacramentum commendantem in Christo et in Ecclesia* (Eng. trans. by John Hammond Taylor, in *The Literal Meaning of Genesis*, Ancient Christian Writers vol. 41 (Paulist Press, 1982), p. 19).

[44] Ibidem, 5, 11: *Quia sive aquae nomine appellare voluit totam corporalem materiam, ut eo modo insinuaret unde facta et formata sint omnia, quae in suis generibus iam dignoscere possumus, appellata aquam, quia ex humida natura videmus omnia in terra per species varias formari atque concrescere; sive spiritalem vitam quamdam ante formam conversionis quasi fluitantem; superferebatur utique Spiritus Dei; quia subiacebat scilicet bonae voluntati Creatoris, quidquid illud erat quod formandum perficiendumque inchoaverat: ut dicente Deo in Verbo suo:* Fiat lux; *in bona voluntate, hoc est in beneplacito eius pro modulo sui generis maneret quod factum est; et ideo rectum est, quod placuerit Deo, Scriptura dicente:* Et facta est lux; et vidit Deus lucem quia bona est (Eng. trans. by John Hammond Taylor, op. cit., p. 25).

Is the separation of light from darkness a marking off of formed creatures from the unformed? And are the terms 'day' and 'night' used to indicate an orderly arrangement, showing that God leaves nothing in disarray, and that the unformed state through which things temporarily pass as they change from form to form is not unplanned? And does this expression imply that the wasting and growth by which creatures succeed one another in the course of time is something that contributes to the beauty of the world? Night certainly consists in darkness which is well ordered".[45]

At last, he suggests proceeding with extreme caution when maintaining an interpretation and discussing it with non-believers: "In matters that are obscure and far beyond our vision, even in such as we may find treated in Holy Scripture, different interpretations are sometimes possible without prejudice to the faith we have received. In such a case, we should not rush in headlong and so firmly take our stand on one side that, if further progress in the search of truth justly undermines this position, we too fall with it. That would be to battle not for the teaching of Holy Scripture but for our own, wishing its teaching to conform to ours, whereas we ought to wish ours to conform to that of Sacred Scripture. ... Usually, even a non-Christian knows something about the earth, the heavens, and the other elements of this world, about the motion and orbit of the stars and even their size and relative positions, about the predictable eclipses of the sun and moon, the cycles of the years and the seasons, about the kinds of animals, shrubs, stones, and so forth, and this knowledge he holds to as being certain from reason and experience. Now, it is a disgraceful and dangerous thing for an infidel to hear a Christian, presumably giving the meaning of Holy Scripture, talking nonsense on these topics; and we should take all means to prevent such an embarrassing situation, in which people show up vast ignorance in a Christian and laugh it to scorn. The shame is not so much that an ignorant individual is derided, but that people outside the household of the faith think our sacred writers held such opinions, and, to the great loss of those for whose salvation we toil, the writers of our Scripture are criticized and rejected as unlearned men. If they find a Christian mistaken in a field which they themselves know well and hear him maintaining his foolish opinions about our books, how are they going to believe those books in matters concerning the resurrection of the dead, the hope of eternal life, and the kingdom of heaven, when they think their pages are full of falsehoods on facts which they themselves have learnt from experience and the light of reason? Reckless and incompetent expounders of Holy Scripture bring untold trouble and sorrow on their wiser brethren when they are caught in one of their mischievous false opinions and are taken to task by those who are not bound by the authority of our sacred books. For then, to defend their utterly foolish and obviously untrue statements, they will try to call upon Holy Scripture for proof and even recite from memory many passages which they think support their position, although *they understand neither what they say nor the things about which they make assertion*".[46]

---

[45] Ibidem, 17, 34: *Sed haec allegoriae propheticae disputatio est, quam non isto sermone suscepimus. Instituimus enim de Scripturis nunc loqui secundum proprietatem rerum gestarum, non secundum aenigmata futurarum. Ergo ad rationem factarum conditarumque naturarum, quomodo invenimus vesperam et mane in luce spiritali? An divisio quidem lucis a tenebris, distinctio est iam rei formatae ab informi; appellatio vero diei et noctis, insinuatio distributionis est, qua significetur nihil Deum inordinatum relinquere, atque ipsam informitatem, per quam res de specie in speciem modo quodam transeundo mutantur, non esse indispositam; neque defectus profectusque creaturae, quibus sibimet temporalia quaeque succedunt, sine supplemento esse decoris universi? Nox enim ordinatae sunt tenebrae* (Eng. trans. by John Hammond Taylor, op. cit., p. 39).

[46] Ibidem, 18.37, 19.39: *Et in rebus obscuris atque a nostris oculis remotissimis, si qua inde scripta etiam divina legerimus, quae possint salva fide qua imbuimur, alias atque alias parere sententias; in nullam earum nos praecipiti affirmatione ita proiciamus, ut si forte diligentius discussa veritas eam recte labefactaverit, corruamus: non pro sententia divinarum Scripturarum, sed pro nostra ita dimicantes, ut eam velimus Scripturarum esse, quae*

He concludes: "Someone will say: 'What have you brought out with all the threshing of this treatise? What kernel have you revealed? What have you winnowed? Why does everything seem to be hidden under questions? Adopt one of the many interpretations which you maintained were possible.' To such a one my answer is that I have arrived at a nourishing kernel in that I have learnt that a man is not in any difficulty in making a reply according to his faith which he ought to make to those who try to defame our Holy Scripture. When they are able, from reliable evidence, to prove some fact of physical science, we shall show that it is not contrary to our Scripture. But when they produce from any of their books a theory contrary to Scripture, and therefore contrary to the Catholic faith, either we shall have some ability to demonstrate that it is absolutely false, or at least we ourselves will hold it so without any shadow of a doubt. And we will so cling to our Mediator, *in whom are hidden all the treasures of wisdom and knowledge,* that we will not be led astray by the glib talk of false philosophy or frightened by the superstition of false religion. When we read the inspired books in the light of this wide variety of true doctrines which are drawn from a few words and founded on the firm basis of Catholic belief, let us choose that one which appears as certainly the meaning intended by the author. But if this is not clear, then at least we should choose an interpretation in keeping with the context of Scripture and in harmony with our faith. But if the meaning cannot be studied and judged by the context of Scripture, at least we should choose only that which our faith demands. For it is one thing to fail to recognize the primary meaning of the writer, and another to depart from the norms of religious belief. If both these difficulties are avoided, the reader gets full profit from his reading. Failing that, even though the writer's intention is uncertain, one will find it useful to extract an interpretation in harmony with our faith".[47]

---

nostra est; cum potius eam quae Scripturarum est, nostram esse velle debeamus. ... Plerumque enim accidit ut aliquid de terra, de coelo, de caeteris mundi huius elementis, de motu et conversione vel etiam magnitudine et intervallis siderum, de certis defectibus solis ac lunae, de circuitibus annorum et temporum, de naturis animalium, fruticum, lapidum, atque huiusmodi caeteris, etiam non christianus ita noverit, ut certissima ratione vel experientia teneat. Turpe est autem nimis et perniciosum ac maxime cavendum, ut christianum de his rebus quasi secundum christianas Litteras loquentem, ita delirare audiat, ut, quemadmodum dicitur, toto coelo errare conspiciens, risum tenere vix possit. Et non tam molestum est, quod errans homo deridetur, sed quod auctores nostri ab eis qui foris sunt, talia sensisse creduntur, et cum magno eorum exitio de quorum salute satagimus, tamquam indocti reprehenduntur atque respuuntur. Cum enim quemquam de numero Christianorum in ea re quam optime norunt, errare comprehenderint, et vanam sententiam suam de nostris Libris asserere; quo pacto illis Libris creditori sunt, de resurrectione mortuorum, et de spe vitae aeternae, regnoque coelorum, quando de his rebus quas iam experiri, vel indubitatis numeris percipere potuerunt, fallaciter putaverint esse conscriptos? Quid enim molestiae tristitiaeque ingerant prudentibus fratribus temerarii praesumptores, satis dici non potest, cum si quando de prava et falsa opinatione sua reprehendi, et convinci coeperint ab eis qui nostrorum Librorum auctoritate non tenentur, ad defendendum id quod levissima temeritate et apertissima falsitate dixerunt, eosdem Libros sanctos, unde id probent, proferre conantur, vel etiam memoriter, quae ad testimonium valere arbitrantur, multa inde verba pronuntiant, non intellegentes neque quae loquuntur, neque de quibus affirmant (Eng. trans. by John Hammond Taylor, op. cit., p. 43).

[47] Ibidem, 21.41: *Dicet aliquis: Quid tu tante tritura dissertationis huius, quid granorum exuisti? quid eventilasti? Cur propemodum in quaestionibus adhuc latent omnia? Affirma aliquid eorum quae multa posse intellegi disputasti. Cui respondeo, ad eum ipsum me cibum suaviter pervenisse, quo didici non haerere homini in respondendo secundum fidem, quod respondendum est hominibus qui calumniari Libris nostrae salutis affectant; ut quidquid ipsi de natura rerum veracibus documentis demonstrare potuerint, ostendamus nostris Litteris non esse contrarium. Quidquid autem de quibuslibet suis voluminibus his nostris Litteris, id est catholicae fidei contrarium protulerint, aut aliqua etiam facilitate ostendamus, aut nulla dubitatione credamus esse falsissimum: atque ita teneamus Mediatorem nostrum, in quo sunt omnes thesauri sapientiae atque scientiae absconditi, ut neque falsae philosophiae loquacitate seducamur, neque falsae religionis superstitione terreamur. Et cum divinos Libros legimus in tanta multitudine verorum intellectuum, qui de paucis verbis eruuntur, et salutate catholicae fidei muniuntur, id potissimum deligamus, quod certum apparuerit cum sensisse quem legimus; si autem hoc latet, id certe quod circumstantia Scripturae non impedit, et cum sana fide concordat: si autem et Scripturae circumstantia*

The subject we are interested in is tackled at the beginning of the second book, where Augustine starts by recommending not confuting those who maintain that the waters cannot stay above the firmament by appealing to the power of God: "Many hold that the waters mentioned in this place cannot be above the starry heaven, maintaining that they would be compelled by their weight to flow down upon the earth or would move in a vaporous state in the air near the earth. No one should argue against this theory by appealing to the power of God, to whom all is possible, and saying that all ought to believe that water, even though it had the same weight as the water we know by experience, was poured forth over the region of the heavens in which the stars are set. For now it is our business to seek in the account of Holy Scripture how God made the universe, not what He might produce in nature or from nature by His miraculous power. If God ever wished oil to remain under water, it would do so. But we should not thereby be ignorant of the nature of oil: we should still know that it is so constituted as to tend towards its proper place and, even when poured under water, to make its way up and settle on the surface. Now we are seeking to know whether the Creator, who has *ordered all things in measure, and number, and weight,* has assigned to the mass of waters not just one proper place around the earth, but another also above the heavens, a region which has been spread around and established beyond the limits of the air. Those who deny this theory base their argument on the weights of the elements. Surely, they say, there is no solid heaven laid out above like a pavement to serve as a support for the mass of water. Such a solid body, they argue, cannot exist except on the earth, and whatever is so constituted is earth, not heaven. They go on to show that the elements are distinguished not by their locations only but also by their qualities, and that each is assigned its place in keeping with its particular qualities".[48]

He then continues by saying that one must not argue with those who discourse of the weight of the elements by using argumentations of the Bible, since by doing so they more readily scorn the sacred books. To favour an *ad litteram* interpretation of the biblical narration, he then speaks at length about descriptions of natural phenomena narrated with everyday language. Nevertheless, he concludes by invoking the authority of the Scripture, which overrides all, "Certain writers, even among those of our faith, attempt to refute those who say that the relative weights of the elements make it impossible for water to exist above the starry heaven. They base their arguments on the properties and motions of the stars. They say that the star called

---

pertractari ac discuti non potest, saltem id solum quod fides sana praescribit. Aliud est enim quid potissimum scriptor senserit non dignoscere, aliud autem a regula pietatis errare. Si utrumque vitetur, perfecte se habet fructus legentis: si vero utrumque vitari non potest, etiam si voluntas scriptoris incerta sit sanae fidei congruam non inutile est eruisse sententiam (Eng. trans. by John Hammond Taylor, op. cit., p. 44).

[48] Ibidem, II, 1.2-1.3: *Multi enim asserunt istarum aquarum naturam super sidereum coelum esse non posse, quod sic habeant ordinatum pondus suum, ut vel super terras fluitent, vel in aere terris proximo vaporaliter ferantur. Neque quisquam istos debet ita refellere, ut dicat secundum omnipotentiam Dei, cui cuncta possibilia sunt, oportere nos credere, aquas etiam tam graves, quam novimus atque sentimus, coelesti corpori, in quo sunt sidera, superfusas. Nunc enim quemadmodum Deus instituerit naturas rerum, secundum Scripturas eius nos convenit quaerere; non quid in eis vel ex eis ad miraculum potentiae suae velit operari. Neque enim si vellet Deus sub aqua oleum aliquando manere, non fieret; non ex eo tamen olei natura nobis esset incognita, quod ita facta sit, ut appetendo suum locum, etiam si subterfusa fuerit, perrumpat aquas, eisque se superpositam collocet. Nunc ergo quaerimus utrum conditor rerum, qui omnia in mensura et numero et pendere disposuit, non unum locum proprium ponderi aquarum circa terram tribuerit, sed et super coelum quod ultra limitem aeris circumfusum atque solidatum est. Quod qui negant esse credendum, de ponderibus elementorum argumentantur, negantes ullo modo ita desuper quasi quodam pavimento solidatum esse coelum, ut possit aquarum pondera sustinere; quod talis soliditas nisi terris esse non possit, et quidquid tale est, non coelum sed terra sit. Non enim tantum locis, sed etiam qualitatibus elementa distingui, ut pro qualitatibus propriis etiam loca propria sortirentur* (Eng. trans. by John Hammond Taylor, op. cit., pp. 46-47).

Saturn is the coldest star, and that it takes thirty years to complete its orbit in the heavens because it is higher up and therefore travels over a wider course. The sun completes a similar orbit in a year, and the moon in a month, requiring a briefer time, they explain, because these bodies are lower in the heavens; and thus the extent of time is in proportion to the extent of space. These writers are then asked why Saturn is cold. Its temperature should be higher in proportion to the rapid movement it has by reason of its height in the heavens. For surely when a round mass is rotated, the parts near the center move more slowly, and those near the edge more rapidly, so that the greater and lesser distances may be covered simultaneously in the same circular motion. Now, the greater the speed of an object, the greater its heat. Accordingly, this star ought to be hot rather than cold. It is true, indeed, that by its own motion, moving over a vast space, it takes thirty years to complete its orbit; yet by the motion of the heavens it is rotated rapidly in the opposite direction and must daily travel this course (and thus, they say, each revolution of the heavens accounts for a single day); and, therefore, it ought to generate greater heat by reason of its greater velocity. The conclusion is, then, that it is cooled by the waters that are near it above the heavens, although the existence of these waters is denied by those who propose the explanation of the motion of the heavens and the stars that I have briefly outlined. With this reasoning some of our scholars attack the position of those who refuse to believe that there are waters above the heavens while maintaining that the star whose path is in the height of the heavens is cold. Thus they would compel the disbeliever to admit that water is there not in a vaporous state but in the form of ice. But whatever the nature of that water and whatever the manner of its being there, we must not doubt that it does exist in that place. The authority of Scripture in this matter is greater than all human ingenuity".[49]

But the fundamental argument to which one must resort in any case is that Scripture wants to teach men the salvation of their soul and not the scientific truths: "It is also frequently asked what our belief must be about the form and shape of heaven according to Sacred Scripture. Many scholars engage in lengthy discussions on these matters, but the sacred writers with their deeper wisdom have omitted them. Such subjects are of no profit for those who seek beatitude, and, what is worse, they take up very precious time that ought to be given to what is spiritually beneficial. What concern is it of mine whether heaven is like a sphere and the earth is enclosed by it and suspended in the middle of the universe, or whether heaven like a disk above the earth covers it over on one side? But the credibility of Scripture is at stake, and as I have indicated more than once, there is danger that a man uninstructed in divine revelation, discovering

---

[49] Ibidem, 5.9: *Quidam etiam nostri, istos negantes propter pondera elementorum aquas esse posse super coelum sidereum, de ipsorum siderum qualitatibus et meatibus convincere moliuntur. Ildem namque asserunt stellam quam Saturni appellant, esse frigidissimam, eamque per annos triginta signiferum peragere circulum, eo quod superiore ac per hoc ampliore ambitu graditur. Nam sol eumdem circulum per annum complet, et luna per mensem; tanto, ut dicunt, brevius, quanto inferius, ut spatio loci spatium temporis congruat, Quaeritur itaque ab eis, unde illa stella sit frigida, quae tanto ardentior esse deberet, quanto sublimiore coelo rapitur. Nam procul dubio cum rotunda moles circulari motu agitur, interiora eius tardius eunt, exteriora celerius, ut maiora spatia cum brevioribus ad eosdem gyros pariter occurrant: quae autem celerius, utique ferventius. Proinde memorata stella magis debuit calida esse quam frigida: quamvis enim suo motu, quoniam grande spatium est, triginta annis totum ambitum permeet, tamen coeli motu in contrarium rotata velocius, quod quotidie necesse est patiatur (sic, ut dicunt, coeli singulae conversiones, dies singulos explicant), calorem maiorem debuit coelo concitatiore concipere. Nimirum ergo eam frigidam facit aquarum super coelum constitutarum illa vicinitas, quam nolunt credere, qui haec, quae breviter dixi, de motu coeli et siderum disputant. His quidam nostri coniecturis agunt adversus eos qui nolunt aquas super coelum credere, et volunt eam stellam esse frigidam, quae iuxta summum coelum circuit; ut ex hoc cogantur aquarum naturam, non iam illic vaporali tenuitate, sed glaciali soliditate pendere. Quoque modo autem et qualeslibet aquae ibi sint, esse eas ibi minime dubitemus: maior est quippe Scripturae huius auctoritas, quam omnis humani ingenii capacitas* (Eng. trans. by John Hammond Taylor, op. cit., pp. 51-52).

something in Scripture or hearing from it something that seems to be at variance with the knowledge he has acquired, may resolutely withhold his assent in other matters where Scripture presents useful admonitions, narratives, or declarations. Hence, I must say briefly that in the matter of the shape of heaven the sacred writers knew the truth, but that the Spirit of God, who spoke through them, did not wish to teach men these facts that would be of no avail for their salvation. But someone may ask: 'Is not Scripture opposed to those who hold that heaven is spherical, when it says, *who stretches out heaven like a skin?*' Let it be opposed indeed if their statement is false. The truth is rather in what God reveals than in what groping men surmise. But if they are able to establish their doctrine with proofs that cannot be denied, we must show that this statement of Scripture about the skin is not opposed to the truth of their conclusions. If it were, it would be opposed also to Sacred Scripture itself in another passage where it says that heaven is suspended like a vault. For what can be so different and contradictory as a skin stretched out flat and the curved shape of a vault? But if it is necessary, as it surely is, to interpret these two passages so that they are shown not to be contradictory but to be reconcilable, it is also necessary that both of these passages should not contradict the theories that may be supported by true evidence, by which heaven is said to be curved on all sides in the shape of a sphere, provided only that this is proved. Our picture of heaven as a vault, even when taken in a literal sense, does not contradict the theory that heaven is a sphere. We may well believe that in speaking of the shape of heaven Scripture wished to describe that part which is over our heads. If, therefore, it is not a sphere, it is a vault on that side on which it covers the earth; but if it is a sphere, it is a vault all around. But the image of the skin presents a more serious difficulty: we must show that it is reconcilable not with the sphere (for that may be only a man-made theory) but with the vault of Holy Scripture. My allegorical interpretation of this passage can be found in the thirteenth book of my *Confessions*. Whether the description of heaven stretched out like a skin is to be taken as I have interpreted it there or in some other way, here I must take into account the doggedly literal-minded interpreters and say what I think is obvious to everyone from the testimony of the senses. Both the skin and the vault perhaps can be taken as figurative expressions; but how they are to be understood in a literal sense must be explained. If a vault can be not only curved but also flat, a skin surely can be stretched out not only on a flat plane but also in a spherical shape. Thus, for instance, a leather bottle and an inflated ball are both made of skin".[50]

---

[50] Ibidem, 9.20-9.22: *Quaeri etiam solet quae forma et figura coeli esse credenda sit secundum Scripturas nostras. Multi enim multum disputant de iis rebus, quas maiore prudentia nostri auctores omiserunt, ad beatam vitam non profuturas discentibus; et occupantes, quod peius est, multum pretiosa, et rebus salubribus impendenda temporum spatia. Quid enim ad me pertinet, utrum coelum sicut sphaera undique concludat terram in media mundi mole libratam, an eam ex una parte desuper velut discus operiat? Sed quia de fide agitur Scripturarum, propter illam causam, quam non semel commemoravi, ne quisquam eloquia divina non intellegens, cum de his rebus tale aliquid vel invenerit in Libris nostris, vel ex illis audierit, quod perceptis a se rationibus adversari videatur, nullo modo eis caetera utilia monentibus, vel narrantibus, vel pronuntiantibus credat; breviter dicendum est de figura coeli hoc scisse auctores nostros quod veritas habet; sed Spiritum Dei, qui per ipsos loquebatur, noluisse ista docere homines nulli saluti profutura. Sed, ait aliquis, quomodo non est contrarium iis qui figuram sphaerae coelo tribuunt, quod scriptum est in Litteris nostris: Qui extendit coelum sicut pellem? Sit sane contrarium, si falsum est quod illl dicunt: hoc enim verum est quod divina dicit auctoritas, potius quam illud quod humana infirmitas conicit. Sed si forte illud talibus illi documentis probare potuerint, ut dubitari inde non debeat; demonstrandum est hoc quod apud nos de pelle dictum est, veris illis rationibus non esse contrarium: alioquin contrarium erit etiam ipsis in alio loco Scripturis nostris, ubi coelum dicitur velut camera esse suspensum. Quid enim tam diversum et sibimet adversum, quam plana pellis extensio, et camerae curva convexio? Quod si oportet, sictli oportet, haec duo sic intellegere, ut concordare utrumque, nec sibimet repugnare inveniatur; ita oportet etiam utrumlibet horum illis non adversari disputationibus, si eas forte veras certa ratio declaraverit, quibus docetur coelum sphaerae figura undique esse convexum, si tamen probatur. Et illa quidem apud nos camerae similitudo, etiam secundum litteram accepta, non impedit eos qui sphaeram dicunt. Bene quippe creditur secundum eam*

Finally, he closes the second book by posing a question regarding a problem well present in Greek philosophy, perhaps influenced by his interest in Neoplatonic philosophy: "It is often asked whether the bright luminaries of heaven are bodies only or whether they have spirits within them to rule them; and whether, if they have such spirits, they are made living beings by their souls, or whether there is only the presence of spirits without a vital union. This problem is not easy to solve, but I believe that in the course of commenting on the text of Scripture occasions may present themselves on which we may treat the matter according to the rules for interpreting Holy Scripture, presenting some conclusion that may be held, without perhaps demonstrating it as certain. Meanwhile we should always observe that restraint that is proper to a devout and serious person and on an obscure question entertain no rash belief. Otherwise, if the evidence later reveals the explanation, we are likely to despise it because of our attachment to our error, even though this explanation may not be in any way opposed to the sacred writings of the Old or New Testament".[51]

As one can see, in the second part of the excerpt, Augustine repeats the invitation to caution we have already come across. Whoever is familiar with Galileo's writings recognize in this excerpt the one quoted at the beginning of the letter to Madame Christine of Lorraine of 1615,[52] in which Galileo disproved the "literal" interpretations of the Bible. Anyway, we shall reopen the subject further.

Multa quaesita potius quamvis inventa videantur

This sentence, which Augustine perhaps wrote without adding onto it particular intentions, faithfully expresses, in our opinion, the attitude he displayed in commenting on the biblical narration of the second day of the creation. After all, in a way he was engaged in the study of this subject during a period of at least thirty years and, as we have recalled, in at least five works. The answers to the questions he asks himself are not assertive statements, but rather invitations to take the matter further.

The problem of countering the opinions of the non-believers (philosophers or not) is always

---

partem, quae super nos est, de coeli figura Scripturam loqui voluisse. Si ergo sphaera non est, ex una parte camera est, ex qua parte coelum terram contegit: si autem sphaera est, undique camera est. Sed illud quod de pelle dicturn est, magis urget, ne non sphaerae, quod humanum est forte commentum, sed ipsi nostrae camerae adversum sit. Quid autem hinc allegorice senserim: Confessionum *nostrarum liber tertius decimus habet. Sive igitur ita ut ibi posui, sive aliquo alio modo intellegendum sit coelum sicut pellis extentum; propter molestos et nimios exactores expositionis ad litteram, hoc dico, quod, sicut arbitrar, omnium sensibus patet: utrumque enim fortasse, id est et pellis et camera, figurate intellegi potest; utrumque autem ad litteram quomodo possit, videndum est. Si enim camera non solum curva, sed etiam plana recte dicitur; profecto et pellis non solum in planum, verum etiam in rotundum sinum extenditur. Nam et uter sicut et vesica, pellis est* (Eng. trans. by John Hammond Taylor, op. cit., pp. 59-60).

[51] Ibidem, 18.38: *Solet etiam quaeri, utrum caeli luminaria ista conspicua corpora sola sint, an habeant rectores quosdam spiritus suos: et si habent, utrum ab eis etiam vitaliter inspirentur, sicut animantur carnes per animas animalium, an sola sine ulla permixtione praesentia. Quod licet non facile comprehendi possit; arbitror tamen in processu tractandarum Scripturarum opportuniora loca posse occurrere, ubi nobis de hac re, secundum sanctae auctoritatis regulas, etsi non estendere certum aliquid, tamen credere licebit. Nunc autem servata semper moderatione piae gravitatis, nihil credere de re obscura temere debemus; ne forte quod postea veritas patefecerit, quamvis libris sanctis sive Testamenti Veteris sive Novi nudo modo esse possit adversum, tamen propter amorem nostri erroris oderimus. Nunc ad librum operis nostri iam tertium transeamus* (Eng. trans. by John Hammond Taylor, op. cit., p. 73).

[52] Galileo's letter to Maria Cristina, consort of Ferdinand I de' Medici, Grand Duke of Tuscany, was in defense of the Copernican system, first published in Florence in 1615, it is available in *Le Opere di Galileo Galileo*, vol. 5 (Florence, 1895), pp. 307-348.

hanging over him. As is known, he prefers to debate with the Neoplatonists rather than with the Peripatetics. In fact, regarding the former, he says: "With respect, however, to that wherein they agree with us we prefer them to all others namely, concerning the one God, the author of this universe, who is not only above every body, being incorporeal, but also above all souls, being incorruptible, our principle, our light, our good. And though the Christian man, being ignorant of their writings, does not use in disputation words which he has not learned,—not calling that part of philosophy natural (which is the Latin term) or physical (which is the Greek one), which treats of the investigation of nature; or that part rational, or logical, which deals with the question how truth may be discovered; or that part moral, or ethical, which concerns morals, and shows how good is to be sought, and evil to be shunned, he is not, therefore, ignorant that it is from the one true and supremely good God that we have that nature in which we are made in the image of God, and that doctrine by which we know Him and ourselves, and that grace through which, by cleaving to Him, we are blessed. This, therefore, is the cause why we prefer these to all the others, because, while other philosophers have worn out their minds and powers in seeking the causes of things, and endeavoring to discover the right mode of learning and of living, these, by knowing God, have found where resides the cause by which the universe has been constituted, and the light by which truth is to be discovered, and the fountain at which felicity is to be drunk. All philosophers then, who have had these thoughts concerning God, whether Platonists or others, agree with us. But we have thought it better to plead our cause with the Platonists, because their writings are better known. For the Greeks, whose tongue holds the highest place among the languages of the Gentiles, are loud in their praises of these writings; and the Latins, taken with their excellence, or their renown, have studied them more heartily than other writings, and, by translating them into our tongue, have given them greater celebrity and notoriety".[53]

We can comment on the last part of the excerpt by recalling that Augustine did not actually know Greek until his late maturity (at that time Aristotle's works were not yet translated into Latin) and that in any case Aristotle's philosophy did not consider the creation: the world has always existed, instead of having been created at some point. As a confirmation of the Neoplatonical "affinities" of Augustine, we can add the comparison he makes of the creation of the world as narrated in Genesis with what Plato speaks of in the *Timaeus*.[54]

So, Augustine does not accomplish his study with an enunciation to be considered as an article of faith but, so to speak, entrusts the problem to the philosophers of the Scholastics, for

---

[53] Augustine, *De Civitate Dei* VIII, 10.1-10.2: *In quo autem nobis consentiùnt; de uno Deo huius universitatis auctore, qui non solum super omnia corpora est incorporeus, verum etiam super omnes animas incorruptibilis, principium nostrum, lumen nostrum, bonum nostrum, in hoc eos ceteris anteponimus.Nec, si litteras eorum Christianus ìgnorans verbis, quae non didicit, in disputatione non utitur, ut vel naturalem Latine vel physicam Graece appellet eam partem, in qua de naturae inquisitione tractatur, et rationalem sive logicam, in qua quaeritur quonam modo veritas percipi possit, et moralem vel ethicam, in qua de moribus agitur bonorumque finibus appetendis malorumque vitandis, ideo nescit ab uno vero Deo atque optimo et naturanti nobis esse, qua factì ad eius imaginem sumus, et doctrinam, qua eum nosque noverimus, et gratiam, qua illi cohaerendo beati simus. Haec itaque causa est cur istos ceteris praeferamus, quia, cum alii philosophi ingenia sua studiaque contriverint in requirendis rerum causis, et quinam esset modus discendi atque vivendi, isti Deo cognito reppererunt ubi esset et causa constitutae universitatis et lux percipiendae veritatis et fons bibendae felicitatis. Sive ergo isti Platonici sive quicumque alii quarumlibet gentium philosophi de Deo ista sentiunt, nobiscum sentiunt. Sed ideo cum Platonicis magis agere placuit hanc causam, quia eorum sunt litterae notiores. Nam et Graeci, quorum lingua in gentibus praeminet, eas magna praedicatione celebrarunt, et Latini permoti earum vel excellentìa vel gloria, ipsas libentius didicerunt atque in nostrum eloquium transferendo nobiliores clarioresque fecerunt* (Eng. trans. Philip Schaff in *Nicene and Post-Nicene Fathers First Series: Volume II St. Augustine: City of God, Christian Doctrine* (New York: Cosimo Classics, 2007), p. 151).

[54] Ibidem VIII, 11.

whom it will be one of the opportunities in which the faith clashes with the results of natural science. Augustine does everything he can not to exasperate the conflict and, when it is possible, he avoids it by invoking "things which still are not known", or turns to an interpretation which is "almost allegorical". In any case, since he has a general knowledge of the natural philosophy of the Greeks, he very often finds it difficult to reconcile it with the Bible's text.

Chapter 2
The Waters above the Firmament in the Culture of the Middle Ages

2.1 Up to the Eleventh Century

What is conventionally referred to as the Middle Ages is the *historic* period which goes from the end of the fifth century to the end of the fifteenth AD. We shall have to deal with the development undergone by our question over the course of ten centuries. Whilst our excursion among the works of the Fathers of the Church was facilitated by the fact that most of them attended to an elucidation of the Bible and, often Genesis in particular, the research concerning the commentators of subsequent centuries (from the sixth on) appears more complicated.

First of all let us say that the fall of the Roman Empire had been accompanied by a complete disruption of the cultural institutions and a consequent cultural cancellation of entire populations. In the Roman world, as early as the time of the Republic and then throughout the period of the Empire, there had been a great circulation of encyclopedic compilations which popularized the achievements of Greek natural philosophy.[1] Not always were these "first hand" works, that is, directly based on the Greek originals; often they were remakes of popular works of the Hellenistic period. In any case, the Romans had at their disposal encyclopedias which allowed them to gain access to the scientific results until then achieved. Suffice it to quote, among the many, the monumental *Naturalis Historia* by Pliny the Elder. These encyclopedias, or at least those that have not been lost, will be the sources to which the medieval compilers turn to at first, together of course with the Bible.

We shall divide our research on the medieval authors into two parts. In the first we shall deal with the Early Middle Ages (up to the Carolingian age) and in the second with the other centuries (up to the Renaissance). We are aware that this choice may appear arbitrary and conventional, but it is nevertheless useful.

In time, in the so-called barbarian kingdoms which were formed in Europe after the fall of the Roman Empire, the evangelization by the Catholic Church had met with success by creating bishoprics and afterwards monasteries, which became the starting point for a form of cultural life, connected with the diffusion of the Christian religion and its doctrinal elaboration. One such place was Seville within the Visigothic Kingdom of Spain, which in the past had been one of the more radically Latinized regions (it had produced both emperors and great intellectuals). It is from the work of Isidore, bishop of Seville, that our research will start.

2.1.1 Isidore of Seville (560–630)

The biographical notes regarding Isidore are very scanty. He was born around 560 into a noble family, probably coming from Cartagena; the first information about his public life concern his participation in the resumption of the fight against the Arianism in 583. From 601 until his death in 636 he was bishop of Seville, succeeding his older brother Leander.

In his activity as bishop, he chaired several councils, including the Fourth Council of Toledo

---

[1] See William H. Stahl: *Roman Science*, op. cit.

in 633 where (a rare boast for a Spanish bishop) he deplored the forced conversion of Jews to Christianity imposed by King Sisebut (612–621). However, Isidore's activity as a bishop is not his most distinguishing characteristic, even though his pastoral activity was intense enough to lead to his canonization. Rather it was his irrepressible fertility as a writer which made him the most influential and universally read author of the Christian Middle Ages. Isidore can be considered the greatest scholar of his time and to him the revival is due of the studies in the Visigothic Kingdom of Spain. His cultural interests ranged over all branches of knowledge of his time and he left a huge number of writings, which were used in the abbatial and episcopal schools where the clerics were trained.

In his day and even throughout most of the Middle Ages, his work enjoyed a great reputation and authoritativeness. We can add that even Dante Alighieri mentioned Isidore in his *Divina Commedia* (*"l'ardente spiro di Isidoro"*, Paradiso X.131) and, in our day, in 2002, Pope John Paul II designated him patron saint of Internet. Afterwards, the opinion of scholars was no longer so positive. Substantially, he has come to be considered a compiler and not an original author. Let us quote two eminent historians of the last century. Gabriele Pepe wrote that "more than an encyclopedist he is an intrepid copyist",[2] while Lynn Thorndike said, "the work's importance consists chiefly in showing how scanty was the knowledge of the early middle ages".[3] This opinion is also subscribed to by Ernest Brehaut, who wrote a fundamental critical essay on Isidore in which, after having listed the authors from whom Isidore has drawn his ideas, he concludes, "If viewed closely they are a mass of confusion and incoherence".[4]

Nonetheless, one can say that his work was fundamental since it covered all the fields of interest in his time and the information he gave was the most complete. In any case, one must take into account that Isidore came after centuries of compilations by the Roman encyclopedists, in which inner coherence was not the greatest value, since the necessary condition for making a good divulgation is to master the matter to be divulged, and this was the case of neither the Roman compilers nor of Isidore, at least as far as the natural sciences are concerned.

Let us now take into consideration those among his works which can enlighten us about Isidore's opinion on the waters above the firmament. The work for which he had a universal reputation is that with the Latin title *Etymologiae sive Origines*,[5] an encyclopedia consisting of twenty books, where Isidore deals with the whole of learning that he considers essential for his time, ranging from the more serious and challenging subjects (from the liberal arts, to medicine, to theology, to biology, to architecture, etc.) to the most humble. For instance: the first book bears the title *De Grammatica*, and the twentieth *De penu et instrumentis domesticis et rusticis* (Provisions and various implements). The contents are taken from the Latin works of the preceding centuries, both of the Romans and of the Fathers of the Church, and from the Bible. The notions explained are the most heterogeneous, unified only by the technique of founding the knowledge of the etymology of the words (from this the title *Etymologiae*, or Origins), as if the etymon of the name could explain the nature of the named thing. It goes without saying that the etymologies supplied by Isidore were very often based on the culture of that time and

---

[2] Gabriele Pepe, *Il Medioevo barbarico in Italia* (Il Saggiatore, 1967), p. 468.

[3] Lynn Thorndike, *A History of Magic and Experimental Science* (Columbia University Press, 1923), vol. 1, p. 263.

[4] Ernest Brehaut, *An Encyclopedist of the Dark Ages: Isidore of Seville* (Columbia University Press, 1912), p. 48.

[5] See, for instance, *The Etymologies of Isidore of Seville*, Eng. trans. S. A. Barney, W. J. Lewis, J. A. Beach, O. Berghof, with the collaboration of Muriel Hall (Cambridge: Cambridge University Press, 2006).

even then are anything but rigorous. Isodore worked on drafting this work for more than a decade until his death in 636.

In addition to the *Etymologiae*, two other works meet our case. The first is a little work which is a summary of the knowledge about nature handed down from Antiquity, probably written between 612 and 615, bearing the title *De Natura Rerum*.[6] The second bears the title *Quaestiones in Vetus Testamentum, in Genesim*,[7] written before the *Etymologiae*.

Let us start from the *Etymologiae* to see what general ideas on the world are collected in his principal work. The definition of the heavens provides an example of how the technique of the etymologies operated. In this case, Isidore says, we have two possibilities: according to the first the word *caelum* derives from the image of a chased vase (*vas caelatum*), that is, a vase which sparkles due to the figures (in this case, the stars) hammered into its surface; according to the second, the word *caelum* derives from the act of hiding what is over it (*a superiora caelando*).[8] After having recalled the Greek name of the heavens, he finally says, in regard to the firmament: "In the Scriptures the heaven is called *firmamentum* since it is *firmatum*, or supported, by the course of the constellations and by rational and immutable laws".[9] On the Earth, the explanation is longer. We report an excerpt from the beginning of Book XIV (*De Terra et Partibus*):

Chapter 1. On the earth.

1. The earth is placed in the middle region of the universe, being situated like a centre at an equal interval from all parts of heaven; in the singular number it means the whole circle; in the plural separate parts; and reason gives different names for it; for it is called *terra* from the upper part where it suffers attrition; *humus* from the lower and humid part, as for example, under the sea; again *tellus*, because we take its fruits; it is also called *ops* because it brings opulence. It is likewise called *arva*, from ploughing and cultivating.

2. Earth in distinction from water is called dry; since the Scripture says that 'God called the dry land, earth'. For dryness is the natural property of earth. Its dampness it gets by its relation to water. As to its motion (earthquakes) some say it is wind in its hollow parts, the force of which causes it to move.

3. Others say that a generative water moves in the lands, and causes them to strike together, *sicut vas,* as Lucretius says. Others have it that the earth is sponge-shaped, and its fallen parts lying in ruins cause all the upper parts to shake. The yawning of the earth also is caused either by the motion of the lower water, or by frequent thunderings, or by winds bursting out of the hollow parts of the earth.

Chapter 2. On the circle of lands

1. The circle of lands is so called from its roundness, which is like that of a wheel, whence a small wheel is called *orbiculus*. For the Ocean flowing about on all sides encircles its boundaries. It is divided into three parts; of which the first is called Asia; the second, Europe; the third, Africa.

2. These three parts the ancients did not divide equally; for Asia stretches from the South through the East to the North, and Europe from the North to the West, and

---

[6] See the English edition: Isidore of Seville, *On the Nature of Things*. Translated with introduction, notes, and commentary by Calvin B. Kendall and Faith Wallis (Liverpool University Press, 2016).

[7] See Sancti Isidori, Hispaliensis Episcopi, *Mysticorum Expositiones Sacramentorum seu Quaestiones in Vetus Testamentum. In Genesin*, MPL 083, cols. 207-288 (we are unaware of any translation in a modern language).

[8] Isidore, *Etymologiae* XIII, IV, 1.

[9] Isidore, *Etymologiae* XIII, IV, 1: *Caelum autem in Scripturis sanctis ideo firmamentum vocatur, quod sit cursu siderum et ratis legibus fixisque firmatum* (our Eng. trans.).

thence Africa from the West to the South. Whence plainly the two, Europe and Africa, occupy one-half, and Asia alone the other. But the former were made into two parts because the Great Sea enters from the Ocean between them and cuts them apart. Wherefore if you divide the circle of lands into two parts, East and West, Asia will be in one, and in the other, Europe and Africa.[10]

As appears evident, Isidore's Earth is a disc (*sicut rota est*) coincident with the oecumene, of which he enumerates the nature of the parts. The only "cosmographic" statement is that the Earth has a central position in the universe.

Let us see now what he had written, many years earlier, in the work *De Natura Rerum*, in which he also explicitly speaks about the waters above the firmament. He begins by exposing the views of Ambrose and the philosophers (*philosophy mundi*) about the plurality of the heavens, then ending by saying, "God made them not deformed or confused, but each in its place in a rational order. For the heaven of the highest orb extends itself to the proper limits and assures the equal distance of all its points to the center. And in it are placed the virtues of spiritual creatures. God the creator of the world tempered the fiery nature of his heaven with the water, so that the burning of the superior fire would not kindle the inferior elements. Then he solidified the heaven of the lower orb not only by uniformity but by a multiplicity of motions, calling it the firmament because it sustains the higher waters".[11]

And here, as we can see, the higher waters of the verses of Genesis make their appearance. He then continues, "This thought is from Ambrose: 'Wise men of the world say that waters cannot be over the heavens, saying that if the heaven is fire and the nature of the water is not able to mix with it. They add to which saying the orb of the sky is round and voluble and warm and in that voluble circle water can never stand still. For it is necessary that they (waters) flow and glide when it (heaven) is turned from a higher to a lower orb, so for this reason they can

---

[10] Isidore, *Etymologiae* XIV, I, II: *Terra est in media mundi regione posita, omnibus partibus caeli in modum centri aequali intervallo consistens; quae singulari numero totum orbem significat, plurali vero singulas partes. Cuius nomina diversa dat ratio; nam terra dicta a superiori parte, qua teritur; humus ab inferiori vel humida terra, ut sub mari; tellus autem, quia fructus eius tollimus; haec et Ops dicta, eo quod opem fert frugibus; eadem et arva, ab arando et colendo vocata. Proprie autem terra ad distinctionem aquae arida nuncupatur, sicut Scriptura ait: "Quod vocaverit Deus terram aridam". Naturalis enim proprietas siccitas est terris; nam ut humida sit, hoc aquarum affinitate sortitur. Cuius motum alii dicunt ventum esse in concavitate eius, qui motus eam movet. Sallustius: "Venti per cava terrae citatu rupti aliquot montes tumulique sedere". Alii aquam dicunt genetalem in terris moveri, et eas simul concutere, sicut vas, ut dicit Lucretius. Alii σπογγοειδῆ terram volunt, cuius plerumque latentes ruinae superposita cuncta concutiunt. Terrae quoque hiatus aut motu aquae inferioris fit, aut crebris tonitruis, aut de concavis terrae erumpentibus ventis. Orbis a rotunditate circuli dictus, quia sicut rota est; unde brevis etiam rotella orbiculus appellatur. Undique enim Oceanus circumfluens eius in circulo ambit fines. Divisus est autem trifarie: e quibus una pars Asia, altera Europa, tertia Africa nuncupatur. Quas tres partes orbis veteres non aequaliter diviserunt. Nam Asia a meridie per orientem usque ad septentrionem pervenit; Europa vero a septentrione usque ad occidentem, atque inde Africa ab occidente usque ad meridiem. Unde evidenter orbem dimidium duae tenent, Europa et Africa, alium vero dimidium sola Asia; sed ideo istae duae partes factae sunt, quia inter utrumque ab Oceano mare Magnum ingreditur, quod eas intersecat. Quapropter si in duas partes orientis et occidentis orbem dividas, Asia erit in una, in altera vero Europa et Africa.* (Eng. trans. by Ernest Brehaut in *An Encyclopedist of the Dark Ages*, op. cit., pp. 243-244).

[11] Isidore, *De Natura Rerum* XIII, 2: *Fecit autem Deus non informes, vel confusos, sed ratione quadam ordine suo distinctos. Nam superioris circuli coelum proprio discretum termino, et aequalibus undique spatiis collectum ostendit, atque in eo virtutes spiritualium creaturarum constituit. Cujus quidem coeli naturam artifex mundi Deus aquis temperavit, ne conflagratio superioris ignis inferiora elementa succenderet. Dehinc circulum inferioris coeli, non uniformi, sed multiplici motu solidavit, nuncupans eum firmamentum propter sustentationem superiorum aquarum* (Eng. trans. Carolyn Embach, available at: https://www.researchgate.net/publication/315663902_On_the_Nature_of_Things_De_Natura_Rerum_by_Isidor e_of_Seville_ca_560-636_AD_Translated_by_Carolyn_Embach_1969, accessed 12.01.2020).

never stand still there because the axis of the sky turning and revolving pours them forth with swift motion'. But they are being nonsensical and speak confusedly, because Whoever can create something from nothing can set waters in the sky with the nature of solid ice. For when they say the shining orb of the heaven revolves with burning stars, divine providence necessarily looks on, so that between the orb of the heaven water flows and the heat of the burning axis is tempered".[12]

Up to this point there is no sign of a further elaboration with respect to Augustine's *ad litteram* interpretation. Rather there is a return (a regression?) to Lactantius, both regarding the model of oecumene (*sicut rota est*) and the final diatribe against all who are doubtful about the existence of waters above the firmament.

But Isidore does not limit himself to the only *ad litteram* interpretation. Let us see how he tackles the allegorical interpretation in the *Quaestiones in Vetus Testamentum, in Genesim*. As we have recalled above, the writing of this work preceded that of the *Etymologiae* and in contrast to the later work the earlier one is written in a concise style, as the author points out in the brief *Prefatio*. In fact, Isidore says that the work is not only addressed to scholars (*non solum studiosis*) but also to impatient readers (*sed etiam fastidiosis lectoribus*) who detest the talks of excessive length (*qui nimiam longitudinem sermonis abhorrent*). And he continues: "The concise talks do not provoke repulsion for the prolixity. A prolix and obscure speech bores, a concise and open one delights. And since already in the past the discourse has been constructed by us completely to the letter, it is necessary, as a previous basis for the history, that the allegorical interpretation follows. In fact certain things among those are intelligible figuratively, just as the predictive signs of future events".[13] He concludes the *Prefatio* by saying that what he asserts is taken (in his own voice and language) from the sages who have preceded him: Origen, Victorinus, Ambrose, Jerome, Augustine, Fulgentius, Cassianus and Gregory.[14]

The creation of the second day is described in this way: "6. Then in the second day God arranged the firmament, i.e., the consolidation of the Scriptures; in fact in the Church the firmament is interpreted as the Scriptures, so it is written: *the heaven will unfold as a book* (Isa. XXXIV, 4). And above this firmament he divided the waters, that is, the celestial peoples of the angels, who have no need of turning their gaze towards this firmament so that those who read hear the word of God. 7. In fact they see him always, and love him; but he superimposed the same firmament of his law to the weakness of the lower peoples, so those who watch there

---

[12] Isidore, XIV, 1, 2: *Haec est Ambrosii sententia: Aquas super caelos sapientes mundi hujus aiunt esse non posse, dicentes: igneum esse coelum, non posse concordari eum eo naturam aquarum. Addunt quoque, dicentes rotundum, ac volubilem, atque ardentem esse orbem coeli, et in illo volubili circuitu aquas stare nequaquam posse. Nam necesse est, ut defluant, et labantur, eum de superioribus ad inferiora orbis ille detorquetur, ac per hoc nequaquam eas stare posse aiunt, quod axis coeli concito se motu torquens eas volvendo effunderet. Sed hi tandem insanire desinant, atque confusi agnoscant, quia qui potuit cuncta creare ex nihilo, potuit et illam aquarum naturam glaciali soliditate stabilire in coelo. Nam cum et ipsi dicant volvi orbem stellis ardentibus re fulgentem, nonne divina Providentia necessario prospexit, ut inter orbem coeli redundarent aquae, quae illa ferventis axis incendia temperarent?* (Eng. trans. Carolyn Embach, op. cit.).

[13] Isidore, *Quaestiones in Vetus Testamentum*, op. cit., Prefatio, 3 (cols. 207-208): *Brevi enim expositione succinta non faciunt de prolixitate fastidium. Prolixa enim et occulta taedet oratio; brevis et aperta delectat. Et quia jam pridem juxta litteram a nobis sermo totus contextus est, necesse est ut, precedente historiae fundamento, allegoricus sensus sequatur, Nam figuraliter quaedam ex his intelliguntur, vere tamquam prophetica indicia praecedentia futurorum* (our Eng. trans.).

[14] Isidore, *Quaestiones in Vetus Testamentum*, op. cit., Prefatio, 5 (col. 209): *...vox mea, ipsorum est lingua. Sumpta itaque sunt ab auctoribus: Origene, Victorino, Ambrosio, Hieronymo, Augustino, Fulgentio, Cassiano ac nostri temporis insigniter eloquenti Gregorio* (our Eng. trans.).

know how to separate the bodily things from the spiritual ones, as the upper waters from the lower ones".[15]

From the excerpts of Isidore's works reported above, it is not easy to judge if his writings constitute an original contribution regarding the subjects dealt with or if he represented a shared opinion, at least within a certain ambit. Ernest Brehaut has already asked this question: "Is it possible to ascertain from the writings of Isidore what was the general view of the universe and the attitude toward life held in the sixth and seventh centuries?", to which he answered himself, "On first thought it seems doubtful".[16] Then he added in a footnote, "...Again, he never tells us whether he believed the earth to be flat or spherical; he uses at one time language that belongs to the spherical earth, and at another, language that can have sense only if he believed the earth to be flat. Here we have not only no definite statement of the conception—although it must have existed in his mind, considering the frequency of his writings on the physical universe— but we have in addition the puzzle of deciding which set of expressions used in this connection was meaningless to him".[17]

As to the flatness or the sphericity of the Earth, he reminds us more of the unshakable certainty of Lactantius than the painful uncertainty of Augustine. With Augustine he shared not even the uncertainty regarding the prevailing interpretation (that is, when one type of interpretation must prevail over the other) of Genesis. As we have seen, Isidore simply juxtaposes the two interpretations in different works. He expounds them one after the other as if he were scrupulous about not neglecting things which one must necessarily report. It seems to us that we can conclude that Isidore does not hand down to his successors a decision (right or wrong) regarding the problem of the waters above the firmament, and limits himself to listing the possible interpretations without taking a stand, as if they can be both valid at the same time.

2.1.2 Bede the Venerable (672–735)

We can somehow liken to the work of Isidore to that of Bede the Venerable. Bede was born in 672 or 673 (thus about a century after Isidore) in Britain in one of the barbaric kingdoms— to be precise, the Kingdom of Northumbria on the eastern coast—which were established in the beginning of the seventh century. In this case as well this was a region which had been subject to an intense Christian evangelization and to the building of several monasteries.

According to what he wrote about himself in the most renowned of his works (*Historia ecclesiastica gentis Anglorum* V, 24, 2),[18] Bede spent all his life as a monk in the two "twin" monasteries of Wearmouth and Yarrow, entirely devoted to prayer, intense and prolific study and teaching. According to historians, Bede provided English Catholicism with the same intellectual equipment that Isidore supplied to Spanish Catholicism. He wrote many works of history, theology and science, and for this he was considered the most important intellectual of his time. Of course, not all historians agree. Also in this case, Lynn Thorndike was not generous:

---

[15] Isidore, *Quaestiones in Vetus Testamentum*, op. cit., I, 6-7 (col. 210): 6. *Deinde secunda die disposuit Deus firmamentum, id est, solidamentum sanctarum Scripturarum; firmamentum enim in Ecclesia Scripturae divinae intelliguntur, sicut scriptum est:* Caelum plicabitur sicut liber *(Isai. XXXIV, 4). Discrevitque super hoc firmamentum aquas, id est, coelestes populos angelorum, qui non opus habent hoc suspicere firmamentum, ut legentes audiant verbum Dei. 7. Vident enim eum semper, et diligunt,sed superposuit ipsum firmamentum legis suae super infirmitatem inferiorum populorum, ut ibi suspicientes cognoscant qualiter discernant inter carnalia et spiritualia, quasi inter aquas superiores et inferiores* (our Eng. trans.).

[16] E. Brehaut, *An Encyclopedist of the Dark Ages*, op. cit., p. 27.

[17] Ibidem.

[18] See, for instance, Bede, *The ecclesiastical history of the English people. The Greater Chronicle, Bede's letter to Egbert* (Oxford World's Classics, 2008).

"Bede perhaps knew more natural science than anyone of his time, but if so, the others must have known practically nothing; his knowledge can in no sense be called extensive".[19] Thorndike is obviously referring to Bede's works concerning the divulgation of natural sciences, which can be included in the tradition of the Christian encyclopedists, such as Isidore.

One of the most discussed questions in the Christian field at that time (and for many centuries to come!) was the determination of the date of Easter; the task was to determine a date in the Julian calendar starting from the story told in the Gospel, which was based on references to a lunar calendar. Bede was certainly the most competent author on the subject and tried to explain the thing to his fellows in an accessible way, since he wrote not "as a critic for critics but as a student of sacred literature whose object is the instruction and edification of devote minds", as Claude Jenkins writes.[20]

Bede devoted to this subject a first little work, *De Temporibus*,[21] followed in the maturity by a longer work *De Temporum Ratione*.[22] At almost the same time with the first, he also wrote the little work *De Natura Rerum*,[23] which is the one of interest for us, for obvious reasons. This is a concise booklet comprising 51 chapters in which the fundamental notions on the nature and the structure of the universe are presented. Except for the initial part, it follows Isidore's homonymous work, even while differing from it in certain conclusions. The "scientific" ideas are more often derived from Pliny, where the *Naturalis Historia* prevails over Isidore.

Bede also follows Pliny in his blunders (for instance: the size of the moon in comparison with that of the earth). But, most important for us, Bede follows Pliny as regards the shape of the earth. Since it was a textbook conceived with a teaching intent, this is an important point. A confirmation is given by Lynn Thorndike, an expert in medieval manuscripts, who remarks, "In addition to Bede's own statement of his aim, the frequency with which we find manuscripts of early date of the *Natura Rerum* and *De Temporibus* suggests they were employed as textbooks in the monastic schools of the early middle ages".[24] Therefore, it is reasonable to think that these two little works had a notable impact on the formation of ideas regarding the matters we are dealing with in the only environment in which at that time learning was circulating.

Let us focus now on the *De Natura Rerum*. Although, on one hand, the extreme schematization pursued by Bede did not allow the student monk to broaden his knowledge of a subject, on the other it does allow us to clearly understand Bede's thought regarding certain subjects. Let us consider the three brief chapters which sum up the subject we are interested in:

> 3. What the World is
>   The world is the entire universe, which consists of heaven and earth, rounded out
>   of four elements into the appearance of a complete sphere: out of fire, by which the
>   stars shine; out of air, by which all living things breathe; out of the waters, which
>   barricade the earth by surrounding and penetrating it; and out of earth itself, which

---

[19] L. Thorndike, *A History of Magic and Experimental Science*, op. cit., vol. I, p. 634.

[20] C. Jenkins, "Bede as exegete and theologian", in *Bede: His life, Times and Writings. Essays in Commemoration of the Twelfth Centenary of his Death*, 2nd ed., Alexander Hamilton Thompson, ed. (Oxford Clarendon Press, 1969), p. 170.

[21] See Bede, *De Temporibus liber*, MPL 090, cols. 277-292.

[22] See Bede, *De Temporum ratione*, MPL 090, cols. 293-578.

[23] See Bede, *De Natura Rerum liber*, MPL 090, cols. 187-278. On these three works, see *Bede: On the Nature of things and On Times*, translation with introduction, notes and commentary by Calvin B. Kendall and Faith Wallis (Liverpool University Press, 2010).

[24] L. Thorndike, *History of Magic and Experimental Science*, op. cit., pp. 634-635.

is the middle and lowest portion of the world. It hangs suspended, motionless, with the universe whirling around it. But heaven is also called by the word 'mundus', meaning 'elegant', from its perfect and absolute elegance; for it is called 'cosmos' by the Greeks from its adornment.

...

5. The Firmament

Heaven is of a fine and fiery nature, and round and arranged on all sides at equal distances from the centre of the earth. Hence it appears to be vaulted and centred wherever it may be viewed. Those who are knowledgeable about the world have stated that it revolves daily with indescribable swiftness, so that it would destroy itself if it were not restrained by the countervailing course of the planets.

...

8. The Heavenly Waters

Some people maintain that the waters placed above the firmament, lower indeed than the spiritual heavens but nevertheless superior to every corporeal creation, were reserved for the inundation of the Flood, but others claim more correctly that they were suspended to temper the fire of the stars.[25]

Here, in as concise a form as possible, Bede summarizes the opinions about the waters above the firmament expressed by the Fathers of the Church, from Basil to Ambrose and Augustine. He will be wordier in the work devoted to the comment on the six days of the creation in Genesis,[26] on which he had worked on several occasions before arriving at its final version of four books. As Calvin Kendall says,[27] Bede structured the four books of his commentary on the theme of exile, which in the Middle Ages (for Bede and his contemporaries) was the fundamental metaphor for the condition of human life on Earth. In view of our definite and well defined aim here, it is more important for us to focus on the type of interpretation preferred in the commentary.

In this connection, still referring to the learned introduction with which Calvin Kendall prefaced his translation, Bede also theorized the methodology one must use in commenting on the *Bible*, proceeding to the quest for the spiritual meaning, after having followed the "letter". Obviously, these are arguments (which also concern the problem of harmonizing the Testaments) into which we cannot go.

In order to understand the spirit with which he tackles the narration of the second day of the creation, it will suffice for us to report what Bede says at the beginning of the commentary of the first book: "But it must be carefully observed, as each one devotes his attention to the

---

[25] Bede, *De Natura Rerum* III, V, VIII: *III: Quid sit mundus. Mundus est universitas omnis, quae constti et coelo et terra, quatuor elementis in speciem orbis absoluti globata : igne, quo sidera lucent; aere, qua cuncta viventia spirant; aquis, quae terram cingendo et penetrando communiunt; atque ipsa terra, quae mundi media atque ima, librata volubili circa eam universitate pendet immobilis. Verum mundi nomine etiam eoelum a perfecta absolutaque elegantia vocatur; nam et apud Graecos ab ornata χόσμος appellatur. ... V: De firmamento. Coelum subtilis igneaeque naturae, rotundumque, et a centro terrae aequis spatii undique collectum. Unde et convexum mediumque, quacunque cernatur, inenarrabili celeritate quotidie circumagi sapientes mundi dixerunt, ita ut rueret, si non planetarum occursu moderaretur... VIII: De aquis coelestibus. Aquas, super firmamentum positas, coelis quidem spiritalibus humiliores, sed tamen omni creatura corporali superiores, quidam ad inundationem diluvii servatas, alii vero rectius ad ignem siderum temperandum suspensas ad firmant* (Eng. trans. Calvin B. Kendall and Faith Wallis, op. cit. 2010, p. 75).

[26] Bede, *Hexaemeron sive Libri Quatuor in Principium Genesis usque ad nativitatem Isaac et electionem Ismaelis*, MPL 091, cols. 10-188. See the recent translation: Bede, *On Genesis*, trans. and commentary Calvin Kendall (Liverpool University Press, 2008).

[27] Bede, *On Genesis*, op. cit., p.14.

allegorical senses how far he may have forsaken the manifest truth of history by allegorical interpretation".[28] Then Bede comments on the verses which narrate the second day of the creation in this way: "Here is described the creation of our heaven in which are the fixed stars. It is certain that this firmament is in the midst of the waters, for we ourselves see the waters that were placed beneath it and in the air and lands, and we are informed about those that were placed above it, not only by the authority of this Scriptural passage, but also by the words of the Prophet, who says, *stretching out the heaven like an animal skin, you cover its upper parts with waters.* Therefore it is known that the starry heaven was created in the midst of the waters, nor does anything prevent a belief that it was also made from the waters. For what prevents us, who know how great is the firmness as well as the transparency and purity *of crystalline rock,* which is known to have been made from *the congealing of waters,* from believing that the same Disposer of the things of nature solidified the substance of waters in the firmament of heaven? But if it puzzles anyone, how the waters, whose nature it is always to flow and to sink to the lowest point, can settle above heaven, whose shape seems to be round, he should remember Holy Scripture saying about God, *He binds up the waters in his clouds, so that they break not out and fall down together.* And he should understand that God, who when he wishes and as occasion warrants binds up the waters beneath heaven, which are supported by no foundation of a firmer substance, but are held only by the vapours of the clouds so that they too may not fall, could also suspend the waters above the round sphere of heaven, not with vaporous thinness but with ice-like solidity, so that they would never fall. But although he willed to fix the liquid waters there, is this any greater a miracle than that, as Scripture says, the very *bulk of the earth he has hanged on nothing?* And when the waters both of the Red Sea and of the river Jordan are lifted up on high and made firm like walls for the passage of the people of Israel, do they not give visible evidence that, even beyond the revolving rotundity of heaven, they can remain in a fixed position? Of course, what sort of waters may be there and for what uses they may have been reserved, the Creator himself would know; only he left no room for doubt that they are there, because Holy Scripture said so. What it is to say of God, let this or that created thing be, has already been said above. For he said that it should be, when he arranged for everything to be created through the Word, that is to say, his only-begotten Son, who is coeternal with himself".[29]

---

[28] Bede, *Hexaemeron,* op. cit., col. 13: *Sed diligenter intuendum, ut ita quisque sensibus allegoricis studium impendat, quatenus apertam historiae fidem allegoriando non derelinquat* (Eng. trans. by Calvin Kendall, *On Genesis,* op. cit., p. 69).

[29] Bede, *Hexaemeron,* op. cit., cols. 18-19: *Hic nostri coeli, in quo fixa sunt sidera, creatio describitur ; quod in medio constat firmatum esse aquarum. Nam suppositas esse aquas, et ipsi aere terrisque videmus, superpositas autem non solum hujus. Scripturae auctoritate, sed prophetae verbis edocemur, qui sit:* Extendens coelum sicut pellem, qui tegis in aquis superiore ejus *(Psalm. CIII, 2). In medio ergo aquarum firmatum esse constat sidereum coelum, neque aliquid prohibet ut etiam de aquis factum esse credatur; qui enim cristallini lapidis quanta firmitas, quam sit perspiquitas ac puritas novimus, quem de aquarum concretione certum est esse procreatum, quid obstat credi quod idem dispositor naturarum in firmamento coeli substantiam solidarit aquarum? Si, quem vero movet quomodo aquae, quarum natura est finitare semper atque ad ima delabi, super coelum consistere possint, cujus rotunda videtur esse figura, meminerit Scripturae dicentis de Deo:* Qui ligat aquas in nubibus suis, ut non erumpant pariter deorsum *(Job. XXVI, 8) et intelligat quia qui infra coelum ligat aquas ad tempus cum vult, ut non pariter decidant, nulla firmiora substantiae crepidiae sustentatas, sed vaporibus solum nubium retentas, Ipse etiam potuit aquas super rotundam coeli sphaeram, ne unquam delabantur, non vaporali tenuitate, sed soliditate suspendere glaciali. Sed etsi liquentes, ibi aquas sistere voluit, numquid majoris hoc miraculi est, quam quod ipsam terram molem, ut Scriptura dicit, appendit in nihilo? Nam et undae sive rubri maris, seu fluvii Jordanis, cum ad transitum Israeliticae plebis in altum erectae murorum instar figuntur, nonne evidens dant indicium quod etiam supra rotunditatem coeli volubilem, fixa possint statione manere? Sane quales ibi aquae sint, quosve ad usus reservatae, Conditor ipse noverit; esse tantum eas ibi, qui Scriptura sancta dixit, nulli dubitandum reliquit. Quid sit autem*

As one can see, Bede respects the warning he had given before. His conviction is that the narrated events really happened according to the letter of the text and hence they who expound them must have grasped the literal meaning of the text. It must be taken into account that this work was also written as a guide for the monks who had to preach and so the exposition had to be neat and documented (note the continuous references to Bible verses).

At the beginning of this excerpt, when speaking about the waters, he says that everyone can see the lower ones. Here there is an appeal to experience which causes Kendall to speak of a "protoscientific approach". But the exposition copies what was said by the Fathers of the Church, always in the wake of tradition. When he appeals to divine power for the things unforgivable otherwise, he also adds the fact that the Earth stands still in the universe without being supported by external agents. If we are not mistaken, before that time none of the exegetes had used this argument. Nevertheless, in our opinion Bede does not add new elements to the interpretation of the three verses, apart from a very prudent recourse to the allegorical interpretation.

2.1.3 The Carolingian Renaissance

We are moving along in a quite particular way, since our aim is that of asking the intellectuals who in course of time have dealt with the existence or not of the waters above the firmament. Obviously, within the chronological succession, we make use of a guideline centered on the relevance of the figures taken into consideration. But it may happen that, unlike in the usual historical excursus, a few figures, even important ones, are absent. This is due to the fact that they did not deal with the problem in a specific manner.

Bede the Venerable died in 735 and in the same year Alcuin was born, an intellectual who would be of fundamental importance in the intellectual orientation of Europe for the whole of the Middle Ages. Starting at the end of the eighth century, in Europe Charlemagne (742–814) continually extended his kingdom (of the Franks) by conquering new countries, until he had constructed the Holy Roman Empire (800). In the construction of his Empire, Charlemagne also attended to training what nowadays would be called "the ruling elite", establishing in this way an administrative framework which had to manage not only the administration of the kingdom but also the diffusion of the Catholic Church and its liturgy (even by introducing the Gregorian chants and the Benedictine rule in a new accurate version, the same in all monasteries).

To that end, he founded the Palatine School, devoted to the training those managers and in 781 called Alcuin from England to direct it. Alcuin, at York, had absorbed the tradition of the teaching of the liberal arts and had also become a master of the arts in the episcopal school of York. Called by Charlemagne, he became the organizer of the studies in the Empire of Franks, although he did not live constantly at the court of Charlemagne.

If we wanted to condense into a single sentence the intellectual architecture that was promoted by Alcuin and remained the same throughout the Middle Ages, we should say: the liberal arts as means, the theology as end. This is also the opinion of Émile Brehier, who furthermore points out that Alcuin insisted with strength on the necessity of the liberal arts which he sanctified by showing their bond with the divine creation.[30]

We have mentioned, albeit briefly, the work of Alcuin, since a chain of disciples will

---

dicere Dei fiat haec vel illa creatura, jam supra dictum est. Dixit enim ut fieret, cum in coaeterno sibi Verbo, unigenito videlicet Filio, cuncta creando disposuit (Eng. trans. by Calvin Kendall, On Genesis, op. cit., pp. 75-76).

[30] Émile Bréhier, La philosophie du Moyen Age, II, 2 (Paris: Edition Albin Michel, 1937).

originate with him. We shall deal with some of them and wish to contextualize them properly. One disciple of Alcuin was Rabanus Maurus. He was born in Mainz in 780 (or 784) and when still a child entered the Benedictine Abbey of Fulda in Hesse, where he began his studies. Then he studied in Tours, one of the great centers of the Carolingian Renaissance, under the leadership of Alcuin. Afterwards he came back to Fulda as a director of the school of the abbey. After a pilgrimage to Jerusalem, he was again at Fulda, where later on (in 822) he became abbot.

His activity as abbot and teacher lasted about twenty years. A great number of doctors came out of his school and went to teach in the nearby provinces. He took care above all of education and for this he also was called *preceptor Germaniae*. His literary production was extensive and his works are in most cases traditional compilations. He dealt, in particular, with the education of the clergy. Regarding this, in his *De clericorum institutione*, a treatise comprising three books,[31] he justifies the use of the profane culture (the liberal arts) with the curious theory of the unjust possession: "About the seven liberal arts of the philosophers, which the Catholics must learn, we have told enough, I think. Eventually we add that, if by chance those who are called philosophers have told true things and in accordance with our faith in their pronouncements or in their writings, above all the Platonists, they must not be feared but claimed for our use from unjust possessors".[32]

In 842, he resigned as abbot and retreated to Petersberg, near Fulda. In 847 he was elected bishop of Mainz and there he died in 856. Leaving aside his activity as a cultural operator in Germany in the Early Middle Ages, we are interested in his activity as an exegete of the Bible, and of Genesis in particular. He wrote a *Commentarium in Genesim Libri quatuor*[33] in the year 819. In this book, obviously, he comments on the verses we are interested in (cols. 449-450), but the commentary is only a copy of Bede's commentary, we have already seen, but of course, following the habit of that time, without mentioning him by name. In this case at least, Rabanus came off with a homage to tradition.

Going further down the chain of the disciples of Alcuin mentioned above, another disciple of Alcuin was Lupus Servatus (805–862), master of Heirich of Auxerre (841–876), in turn master of Remigius of Auxerre (841–908) whom we shall deal with now.

The life of Remigius of Auxerre (Remigius Autissiodorensis) is quite similar to that of other Benedictine monks of the Carolingian age. Born in all probability in Burgundy in 841, when young he entered the Abbey of Saint Germain of Auxerre, where Heirich of Auxerre (a disciple of John Scotus Eriugena) was teacher. Remigius was first a disciple of Heirich and then his successor at his death in 876. He left Auxerre and went to Reims, where he taught until 800, when he moved to Paris, There he died in 908. Like his contemporaries, in the cultural context of the Carolingian Renaissance he was both humanist and theologian. A great expert of Latin, it seems he also knew Greek. In any case, he wrote many commentaries on Latin authors and on the Bible. He is one of the most important among the ranks of the exegetes philologists.

As regards the subject we are interested in, we shall refer to the work *Commentarius in*

---

[31] Rabanus Maurus, *De clericorum institutione libri tres*, MPL 107, cols. 293-420A. A recent English translation of this work does not exist.

[32] Rabanus Maurus, *De clericorum institutione libri tres*, col. 404: *Ecce de septem liberalibus artibus philosophorum, ad quam utilitatem discendae sint catholicis, satis, ut reor, superius diximus. Illud adhuc adjicimus, quod philosophi ipsi qui vocantur, si qua forte vera et fidei nostrae accomodata in dispensationibus suis seu scripta dixerunt, maxime Platonici, non solum formidanda non sunt, sed ab eis etiam tamquam injustis possessoribus in usum nostrum vindicanda* (our Eng. trans.).

[33] Rabanus Maurus, *Commentarium in Genesim libri quatuor*, MPL 107, cols. 439-670 B. An English translation does not exist.

Genesim.[34] Let us see, first of all, how Remigius introduces it: "At the beginning of this work are confuted the philosophers who tried to discuss the creation of the world, such as Plato, who maintained that there are three principles—God, clearly, idea and matter—and that God has not created anything from nothing as a author, but has helped in the work of creation of matter. Whereas Aristotle said that the principles are two, matter and form, and the third—I don't know what he wanted to say—was called becoming. One says that Moses was elected by God to demonstrate the truth, remedying the error of those by showing that God has created all things from nothing in the same time, saying not in time but at the beginning of time: *At the beginning*, and understood: *of time*; it is clear without any doubt that time took to itself what before was nothing".[35]

About the heaven, he writes: "*He created* in fact *the heaven and the earth*. We must not interpret it as this visible firmament, but as that empyrean, that is, fiery, or better the intellectual heaven, which is not called fiery for the heat but for the brightness, which is steadily full of angelic spirits, about whom is said in Job: *When the morning stars sang together and all the sons of God shouted for joy* (Job XXXVIII, 7).[36] And here the three known elements are mentioned. In fact, with the name of heaven the air is comprised, with the name of earth the earth itself and the fire which lies hidden in its bowels. The fourth element, indeed, will be mentioned afterwards".[37]

About the waters, he will be very concise: "In that firmament it is clear that the upper waters are divided from the lower waters, and that the upper part is congealed, shaped like a crystalline stone; the other lower is made over into sea. And this was done in the second day".[38]

As one can see from these excerpts, Remigius, like his predecessors (starting from the Pseudo-Hippolytus of five centuries before), is obsessed with the inheritance of the Greek philosophers. Plato is obviously the most studied philosopher (from the scanty works known at that time), given his influence over Augustine, whereas Aristotle is known only second- (or third-!) hand. In any case, among the intellectuals of the Carolingian Renaissance, it was established, with suitable justifications (see, above, Rabanus's theory of the "unjust possessors"), what to accept or refuse. Aristotle, who talks about a world which has no

---

[34] Remigius of Auxerre, *Commentarius in Genesim*, MPL 131, cols. 51-134. A recent edition: *Remigii Autissiodorensis Expositio super Genesim*, ed. Burton Van Name Edwards (Brepols, 1999). There is no complete English translation. A partial translation (or, better, a paraphrase) can be found in Joy A. Schroder, *The Bible in Medieval Tradition: The Book of Genesis* (Wm. B. Edermans Publishing, 2015).

[35] Remigius of Auxerre, *Commentarius in Genesim*, op. cit., cols. 53-54: *In principio hujus voluminis philosophi confutantur, qui de mundi creatione conati disputare, sicut Plato, qui tria dixit esse principia, Deus videlicet, exemplar et materiam, et Deum non quasi auctorem ex nihilo cuncta creasse, sed quasi opifici in rebus creandis materiam adjutorium praestitisse. Aristotles autem duo dixit esse principia, materiam scilicet et speciem, tertium qiddam nescio quid volens dicere operatorium appellavit. Quorum erroribus obvians Moyses utpote a Deo electus veritatem dicitur demonstrasse, ostendens Deum cuncta simul ex nihilo formasse, non quidem in tempore, sed in initio temporis dicens:* In principio, *et subauditur:* temporis, *patet procul dubio tunc tempus coepisse quod antea minime erat* (our Eng. trans.).

[36] We report here the English version of the correct Latin text. Remigius's quotation is slightly altered.

[37] Remigius of Auxerre, *Commentarius in Genesim*, op. cit., cols. 54-55: Creavit *enim coelum et terram. Coelum non istud visibile firmamentum accipere debemus, sed illud empyreum, id est igneum, vel intellectuale coelum quod non ab ardore sed a splendore igneum dicitur, quod statim repletum est angelicis spiritibus de quibus in Job dicitur: Cum me laudarent simul astra matutina et jubilarent omnes filii Dei (Job XXXVIII, 7). Et nota tria hic elementa commemorari. Nomine enim coeli aerem collige, nomine terrae ipsam terram, et ignem qui in ejus visceribus latet. Quarti vero elementi, id est aquae, in sequentibus fit commemoration* (our Eng. trans.).

[38] Ibidem, col. 56: *Quo scilicet firmamento aquae superiores divisae sunt ab aquis inferioribus, et illa pars superior in modum crystallini lapidis congelata est, caetera vero inferior in mare redacta est. Et hoc quidem secunda die actum est* (our Eng. trans.).

beginning and then denies the creation, cannot be taken into consideration; but the four Empedoclean elements remain.

Also for the creation, a kind of protocol is followed to the letter by all: in the commentaries one must make a collage of excerpts taken from the Fathers of the Church. We have seen that, regarding the problem of the waters above the firmament, Rabanus Maurus copies Bede and finally Remigius, in a very few lines, simplifies all things: there is no problem. This intellectual community, of which Alcuin was, so to speak, the founding father, has not doubts about the existence of the waters above the firmament. Alcuin himself, who wrote a sort of handbook for the clerics based on questions and answers about Genesis,[39] makes the 281 questions initiate from the seventh day, that is, it was assumed that the faithful would have posed no question about the creation. A position neatly different from that of this group is that of the contemporaneous John Scotus Eriugena (815–877), who also held a position of great prestige at the court of Charles the Bald (823–877), the last successor of Charlemagne.

2.1.4 John Scotus Eriugena (815–877)

John Scotus was born in Ireland, as testified by the two appellatives Scotus and Eriugena (both meaning Irish), around 815, but nothing is known about his youth. In 843, called by the emperor Charles the Bald, he moved to France to head the Palatine School. He was also given the job of translating from the Greek the work of Pseudo-Dionysius the Aeropagite, which Louis the Pious (778–840) had received as a gift from the emperor of Byzantium. The *Corpus Areopagiticum* had already been translated before, but the translation was deemed unsatisfactory. The translation and the commentary of Dionysius' treatises brought John near to Neoplatonism. The authors who exerted a strong influence on him, in addition to Augustine, were the eastern Fathers Basil the Great, Gregory of Nyssa and Maximus the Confessor. Apart from Christian philosophers (and theologians) who influenced him, it is the unanimous opinion of the historians of philosophy that John Scotus is the first true philosopher of the Middle Ages and the forerunner of the Scholastics.

We will not try to summarize what are the cornerstones of his philosophy, but will limit ourselves to pointing out that with him the problem of reconciling the faith with reason (*fides/ratio*) officially comes into being. "*Nisi credideritis non intelligetis*" (Isaiah VII.9), which he often repeated, seems to have been the program of his life: reason has its basis in faith and cannot do anything other than confirm it. Scholars point out that his conception of confirmation of the faith thanks to the reason has often been misunderstood. As a matter of fact, John Scotus was not in a quest for a free-thinking, although during his life his thought was repeatedly regarded as heresy.

His fundamental work was *Periphyseon* (or *De Divisione naturae*),[40] a treatise in five books which he wrote in the last period, about the fundamental theme of the way in which human nature, which came from God as a beginning, would come back to Him as an end? The work is written in the form of a dialogue between a master (*nutritor*) and a disciple (*alumnus*). It is a true dialogue, not a sequence of questions and answers, and the two protagonists are represented as having different characters with different aspects. According to the critics, the *Periphyseon*

---

[39] Alcuin, *Opuscolum Primum. Interrogationes et Responsiones in Genesin*, MPL 100, cols. 515-570.

[40] John Scotus, *Periphyseon, id est De Divisione Naturae Libri Quinque*, MPL 122, cols. 439-1022d. A complete English edition does not exist. The book *John the Scot: Periphyseon*, translated by Myra L. Uhlfelder with summaries by Jean A. Potter (Wipf and Stok Publishers, 2011), is not an unabridged translation, since some parts are summarized.

also has a literary value and is considered perhaps the most important work of medieval thought before the *Summa* by Saint Thomas Aquinas.

In the third book of *Periphyseon*, the theme of creation also is treated, starting from the verses of Genesis, and then also the problem of the waters above the firmament. The space devoted to the second day of creation is rather ample (cols. 693c-698b in the *Patrologia Latina* edition) and we cannot report the excerpt in full. However we shall report that part which contains the assertions we want to comment on:

Master. Let's pass on then to consider the second day. And, first of all, let's say that it is not our intention to undertake an allegorical interpretation according to the moral sense, but we only try to discourse on the creation of the real things in a few words and according to the historical sense under the leadership of God.

Disciple. After all, I don't enquire into this; the Holy Fathers have already supplied allegorical interpretation enough.

Master. *God also said: Let the firmament be made in the midst of the waters and divide the waters from the waters, and God created the firmament.* As far as the firmament is concerned, all unanimously agree to denote as firmament nothing less than this visible heaven. Nevertheless others think that the firmament is that supreme rotating sphere, which wraps round the world from everywhere, and is only adorned by the dances of the stars; others instead think that the firmament is all the space above the moon, where the bodies of the planets are and their course occurs together with the extreme space of the stars; others assert that with the single word firmament one must comprise all the empty space which rotates around the earth, that is, the air and the ether and the highest sphere. In fact they say that, according to the Scriptures, the existence of the air and of the ether cannot be elsewhere. Therefore each one has explained that term in his way; some just for holding up the upper waters, as if physical waters might be above it; others for holding up stellar constellations, which are heavy bodies; others because it contains and consolidates all the visible world. There are even those who want that with that name be called more properly the space of this heavier air, since it holds up, in so far as it can, thanks to a certain bodily consistency of its nature, clouds, rains, thunderstorms, snows, hails and all which rises in it from the vapors of the earth; and, as the whole from the part, from it all other spaces of the upper and lighter visible nature be denominated. I let the readers pass judgement on who really understand it more correctly. It seems to me, considering the meaning of the Greek name στερέωμα, that such a word is suitable since in it the place of all created bodily things is and in it ends. And in fact beyond the firmament nothing sensible, or bodily, or spatial, or temporal is conceivable. Then in it all visible things find their placement. In fact στερέωμα is pronounced almost as στερεα ᾶμα, that is, collected solid bodies: therefore in it all solid things, that is, bodily things, stand and are confined. While about the waters, in whose midst God said that the firmament should be made, I do not easily find something to tell, and not because I do not know what many Holy Fathers thought about them. For instance, Saint Basil in his *Hexaemeron* seems to interpret it in this way: *God on the third day packed and gathered in one place, so that dry land appeared, the lands named abysses, which were scattered all over, outdistanced and shallow, over which at first there was darkness and afterwards the Spirit of God was wandering, and all that after the first light, in the period of three days, had lightened the earth's bulk still unformed. And in the midst of them God said that the firmament should be made.* Saint

Augustine opposes this interpretation, but does not explain sufficiently the nature of the waters within which God put the firmament. In fact, while introducing the opinions of others, he has not clarified his thought and I do not know the reason for this. Anyway, among all, he gives preference to those who maintain that one has to call firmament the spaces of this air which are placed between the sea waters and the underlying river waters and those in the clouds hanging above. And so, not having refuted any interpretation, if you want, I shall briefly explain what I think about these waters.

Disciple. I should like it and it is necessary as well. In fact about this question, it seems to me, no one has been convincing enough.

Master. So I think that every created thing falls within one of these subdivisions: it is wholly Body, or wholly Spirit, or something intermediate between them; since it is neither wholly body, nor wholly spirit but something of intermediate, for a reason of compromise between the extremes it proportionally receives in itself something from the spiritual nature as from the upper extreme, and something from the other nature which is wholly corporeal. For this reason, it exists by itself although sharing in the nature of its extremes. Therefore, anyone who has observed carefully can understand that this world consists of a threefold articulation, since, if it is considered according to the reasons for which it is constituted and substantially exists, it is not only known as spiritual but even wholly as spirit. In fact none of those who correctly philosophize has denied that the reasons for the bodily nature are spiritual, or better, are the spirit itself. But if instead one looks downward, at the lowest parts of the world, that is, at all these bodies made up of the universal elements, chiefly at those terrestrial and aqueous which are attached to generation and corruption, he will not find anything other than body and corporeity. Yet, anyone who turns to look at the nature of the simple elements, will find with extreme clarity a certain mean proportion owing to which they neither are wholly bodily, notwithstanding the natural bodies survive their corruption thanks to coitus, nor are completely devoid of bodily nature, since from them all bodies derive and in them again come back. And again, according to what is said about the reciprocal correspondence with the upper extreme, they neither are wholly spirit since they are not completely released from the corporeity, nor wholly non-spirit since they take from wholly spiritual reasons the occasions of their subsistence. Thus, not unreasonably we said that this world possesses extremes deeply separate from each other and intermediate entities in which the consonant harmony of this universe gathers. Let us say then that the lower parts of this world are the lower waters. And this not without reason, since all which comes up in this world grows and feeds on water; in fact, if one removes the moisture from the bodies, they swiftly run out and diminish and reduce to nearly nothing. In fact the pagan philosophers maintain that also the hottest and burning celestial bodies feed on the aqueous moisture and this is not even denied by the interpreters of the Holy Scriptures. Reason teaches (*ratio edocet*) that the spiritual causes of all which is visible are called upper [celestial] waters. In fact, from these all elements, both

simple or compound, derive as from great sources and are organized in conformity
with intelligible virtues stimulated by them.[41]

---

[41] John Scotus, *Periphyseon*, op. cit., cols. 693C-696A: *N. Transeamus igitur ad secundi diei considerationem. Ac prius dicendum quod de allegoricis intellectibus moralium interpretationum nulla nunc nobis intentio est, sed de sola rerum factarum creatione secundum historiam pauca disserere, deo duce, conamur.*
A. Nec hoc quaero. Satis enim a sanctis patribus de talium allegoria est actum.
*N. Dixit quoque deus: fiat firmamentum in medio aquarum, et diuidat aquas ab aquis. Et fecit deus firmamentum.»
De firmamento omnes unanimiter consentiunt, quod non aliud eo nomine nisi hoc caelum uisibile significatur. Alii tamen extremam illam spheram uolubilem, undique totum mundum ambientem chorisque astrorum ornatam solummodo; alii totum spacium ultra lunam, ubi planetarum corpora et cursus esse creduntur, cum ipso extremo ambito syderum; alii totum inane quod circa terram uoluitur, hoc est aera et aethera sublimissimamque spheram, uno uocabulo firmamenti comprehendi autumant. Non enim alibi legitur, ut aiunt, aeris et aetheris conditio.
Quare autem tali nomine uocatur, prout unicuique uisum est, ita exposuit. Alii quidem propter sustentationem superiorum aquarum, ueluti supra illud corporales aquae sint; alii quia choros siderum sustinet, ueluti quaedam ponderosa corpora; alii quia totum uisibilem mundum intra se contineat ac firmet. Nec desunt qui spacium corpulentioris huius aeris eo nomine proprie uocari uolunt, eo quod nubes, pluuias, imbres, niues, grandines omneque quod ex terrenis uaporibus in eo nascitur firma quadam suae naturae corpulentia sustineat, quantum sustinere potest, ac ueluti totum ex parte cetera spacia leuioris et superioris uisibilis naturae ab eo denominali, Horum uero quis rectius intelligat, legentium arbitrio diiudicandum committo.
Mihi autem greci nominis, quod est στερέωμα, considerata uirtute, tale uocabulum uidetur meruisse, eo quod in eo totius corporalis creaturae situs stet ac terminetur, Vltra nanque firmamentum nil sensibile, uel corporeum, uel locale, uel temporale intelligitur esse. Omnium siquidem uisibilium finis in ipso firmatur. Στερέωμα enim dicitur quasi στερεα ἅμα (hoc est solida simul); in ipso quidem omnia solida (hoc est corporalia) simul terminantur et stant.
De aquis autem, in quarum medio firmamentum fieri deus dixit, non satis reperio quid dicam. Non quod me latuerit quid multi sanctorum patrum de ipsis senserint. Sanctus siquidem Basilius in Examero suo uelle uidetur aquas illas, abyssi nomine uocatas, undique circa terram diffusas rarissimasque et tenuissimas, super quas prius erant tenebrae, deinde spiritus dei superferebatur, et in quibus primitiua lux ueluti trium dierum spaciis giraas informem adhuc terrenam molem resplenduit, densatasque tercia die et congregatas in locum unum ut apparet arida in hoc loco significari, et in earum medio deum dixisse firmamentum fieri. Cui omnino sensui sanctus Augustinus refragatur. Nec tamen de ipsis aquis, intra quas deus firmamentum fecit, satis rationem reddidit. Aliorum nanque opiniones introducens, quid ipse intellexerit - qua occasione, ignoro - non aperuit. Eos autem ceteris praeponit, qui spada huius aeris, quae sunt inter aquas marinas et fluuiales infra se positas et illas in nubibus supra se suspensas, firmamenti nomine uocari contendimi. Nulli itaque sensu refutato, quid de bis aquis senserim, paucis, si libet, rationibus explicabo.
A. Libet quidem, ac ualde est necessarium. De hac enim quaestione adhuc quod mihi satis uideretur a nullo est suasum.
N. Totius itaque conditae naturae trinam diuisionem esse arbitror. Omne enim quod creatum est aut omnino corpus est, aut omnino spiritus, aut aliquod medium, quod nec omnino corpus est nec omnino spiritus, sed quadam medietatis et extremitatum ratione ex spirituali omnino natura, ueluti ex una extremitate et superiori, et ex altera (hoc est ex omnino corporea) proportionaliter in se recipit; unde proprie et connaturaliter extremitatibus suis subsistit. Proinde, si quis intentus inspexerit, in hac ternaria proporrionalitate hunc mundum constitutum intelliget. Siquidem in quantum in rationibus suis, in quibus eternaliter et constitutus est et eternaliter subsistit, consideratur, non solum spiritualis, uerum etiam omnino spiritus cognoscitur. Nemo enim recte philosophantium rationes corporeae naturae spirituales, immo etiam spiritus esse negarit. Dum uero extremae ipsius deorsum uersus inspiciuntur partes, hoc est omnia ista corpora ex catholicis elementis composita, maxime etiam terrena et aquatica, quae et generationi et corruptioni obnoxia sunt, nil aliud in eis inuenitur praeter corpus omnino et corporeum. At si quis simplicium elementorum naturam intueatur, luce clarius quandam proportionabilem medietatem inueniet, qua nec omnino corpus sunt, quamuis eorum corruptione naturalia corpora subsistant et coitu, nec omnino corporeae naturae expertia, dum ab eis omnia corpora profluant et in ea iterum resoluantur. Et iterum alteri superiori quidem extremitati comparat(a) nec omnino spiritus sunt, quoniam non omnino corporea extremitate absoluta, nec omnino non spiritus, cum ex rationibus omnino spiritualibus subsistentiae suae occasiones suscipiant. Non irrationabiliter itaque diximus hunc mundum extremitates quasdam a se inuicem paenitus discretas et medietates, in quibus uniuersitatis concors armonia coniungitur, possidere. Ponamus igitur inferiores huius mundi partes ueluti inferiores aquas. Nec immerito, dum totum quod in hoc mundo nascitur humore crescit atque nutritur. Humida siquidem qualitate corporibus sublata, absque mora tabescunt et*

The author begins with a categorical statement "it is not our intention to undertake an allegorical interpretation according to the moral sense", followed by the resolution of pursuing the historical sense. His intention is strengthened, so to speak, by the answer of the disciple, who says "The Holy Fathers have already supplied allegorical interpretations enough". Therefore it is clear that John wants to carry out his interpretation following the literal narration rationally.

He attacks the subject in a way that is, we might say, circular. In fact, with regard to the waters, he anticipates his idea through an ironical comment. Talking about the firmament, which some think hold the superior waters, he suddenly introduces the comment "as if physical waters might be over it". He has not yet enunciated his interpretation, but anticipates his negative judgement on the physicality of these waters. When he comes to the point, he will say, "about the waters, in whose midst God said that the firmament should be made, I do not easily find something to tell". We interpret this sentence in the sense "there is not much to say", and not in the sense "I don't know what to say". That is, the subject does not require a long explanation.

After having reported Basil's interpretation and, in essence, criticized Augustine because he had not clearly expressed his thought, at the end, as if apologizing, he says "not having refuted any interpretation, if you want, I shall briefly explain what I think about these waters". The disciple reaffirms that none of the explanations given up to then seems to him convincing enough. Then the master, by resorting to the principles of his philosophy regarding nature, obtains as a result the "dematerialization" of the so-called "superior waters". In fact, "Reason teaches that the spiritual causes of all which is visible are called upper waters". In addition, as if what already said were not enough, he goes on to repeat the ironic comment, "There are some who think that above the firmament, that is, above the constellations of the stars, there are very tenuous waters, but the weights and the order of the elements refute them".[42]

The impression one has is that John devotes more arguments (also in that part we did not quote) to establishing the nature of the firmament than that of the upper waters. The final comment we have quoted (the weights and the order of the elements), which one would have expected to be placed in the foreground by who want to interpret the Scripture *ad litteram*, is instead deferred to a "resumption" of the argument which forebodes a digression on the colours and the temperatures of the planets.

In conclusion, we think that he certainly wants to deny the physicality of the upper waters, and then deny their existence as real waters, but in the meantime he does not overly emphasize this conclusion. Caution? After all, John Scotus is the first Christian theologian to deny the existence of "waters" above the firmament which, let us recall, is comprised in a cosmology which is that of the *Timaeus*, which John knew very well.

---

decrescum et pene ad nichilum rediguntur. Nam et caelestia corpora, feruentissima et ignea, humida aquarum natura nutriti sapientes mundi affirmant. Quod nec scripturae sanctae expositores denegant. Spirituales uero omnium uisibilium rationes superioram aquarum nomino appellai! iutlo edocet. Ex ipsis enim omnia dementa, siue simplicia siue composita, ueluti ex quibusdam magnis fontibus defluunt, indeque intelligibili quadam uirtute rigata administrantur (our Eng. trans.).

[42] John Scotus, *Periphyseon*, op. cit., col. 697A: *Sunt qui tenuissimas aquas supra firmamentum (hoc est supra choros siderum) esse putant. Sed eos refellit et ratio ponderum et ordo elementorum* (our Eng. trans.).

2.2 From the Eleventh Century to the Renaissance

In this second part, we shall try to ask the thinkers who, from the year 1000 to the Renaissance, dealt with the waters above the firmament. The typology of the inquiry we have proposed to carry out obliges us to jump centuries between one interlocutor and the next.

Around two centuries after the death of John Scotus, Peter Abelard (1079–1142) was born. Obviously, we cannot summarize the philosophical (and scientific) debate which took place in those two centuries, nor mention prominent personalities such as Gerbert d'Aurillac (we have been unable to find in Gerbert's writings references to the "celestial waters"), nor talk about the "universals", etc.

As we know from history, for Europe the twelfth century was also the century of a new "renaissance",[43] both economic and civil and cultural, and Peter Abelard is undoubtedly the personality who characterizes the philosophical debate in the first half of the twelfth century. Since, from the chronological point of view, after John Scotus he is the first outstanding thinker who deals with the second day of the creation in Genesis, we continue our inquiry starting with him.

2.2.1 Peter Abelard (1079–1142)

Peter Abelard was born at Le Pallet, in the hinterland of Nantes, in 1079 and in the course of his life he moved to different places, first as a disciple and then as a teacher; he founded various schools and had famous disciples. He died in 1142 at Chalon-sur-Sâone. Though nowadays he is also known to a large audience of non-philosophers owing to his dramatic love story with Heloise, in his own time he dominated the scene of the philosophical and theological disputations thanks to his profound learning and rigorous logic. The centre of his personality was the need to research: he wanted to solve on rational grounds every revealed truth, and to address with the tools of dialectic any problem in order to trace it back to an effective human comprehension.

Among his numerous works, there are two that concern our problem. One of them is considered to be his main work: *Sic et Non* (*Yes and No*),[44] in which in a certain sense he codifies his method of research, which afterwards will be the method of the *questions of the Scholastics*. The other one, *Expositio in Hexaemeron*,[45] contains the "non-solution" which Abelard finds for the problem of the "celestial waters".

Abelard's method was to take significant propositions of theology and ethics and to collect from the Fathers of the Church their opinions pro and con, perhaps sharpening the contrast and being careful not to solve any real or apparent contradictions. Abelard only grants infallibility to the Scriptures. He writes, in fact: "If there one finds some apparent contradiction, one must not say '*the author of this book has not kept the truth*', but '*the codex contains mistakes, or the*

---

[43] On this, we limit ourselves to quoting the classical book of Charles Homer Haskins, *The Renaissance of the Twelfth Century* (Harvard University Press, 1927; rpt. Meridian books, the World Publishing Company, 1957).

[44] Peter Abelard, *Sic et Non*, MPL 178, cols. 1329-1610 A. An English translation is *Yes and no: the complete English translation of Peter Abelard's Sic et non*, Patricia Throop, trans. (Charlotte, VT: MedievalMS, 2007. I have not been able to consult this translation personally.

[45] Peter Abelard, *Expositio in Hexaemeron*, MPL 178, cols. 729-784 A. An English translation is Peter Abelard, *An Exposition on the Six-day Work*, Wanda Zemler-Cizewski, trans. (Brepols, 2011). I have not been able to consult this translation personally.

interpreter was wrong, or you do not understand' ".[46] Subsequent authorities may err for other reasons and, when they disagree, he claims the right to interfere with his reasoning to investigate the truth. His belief is: *dubitando enigma ad inquisitionem venimus; inquired veritatem percipimus* ("by doubting we come to inquiry, and by inquiring we perceive the truth"). However, some time ago the historian of medieval philosophy Émile Bréhier warned that Abelard's alleged rationalism is a modern invention.[47]

Let us see him at work in his commentary to Genesis: "*Let a firmament be made amidst the waters and let it divide the waters from the waters*". He calls firmament the airy heaven and, in the like manner and together, the ethereal heaven, of which it is said "*the waters which stand above the heavens*", that is, which stand between these lower waters, that is, the earth, and those upper waters. Therefore this heaven is named firmament since its placement stops the fluid nature of the upper water so that it does not flow and fall down. And that's why Jerome says that the Hebraic term '*Samaim*', that is the airy and the ethereal heavens, which here Moses calls firmament, is a word which has its roots in the waters, and is named in this way because it stops the upper waters so that they do not fall down. One wonders how the fire and the air can hold up the substance of the water which is heavier. But undoubtedly so great may be the fluidity and the lightness of those waters and so great the mass of the air and the fire subjected to it that those are able to be sustained by these elements, as waters hold up wood and stones even if they are of earthy nature and thus heavier. Moreover, who does not know that the air next to the waters, despite being heavier, nevertheless, with the exhalation of the earth, attracts the waters in the form of vapor and hold up them before they set into drops? If then those upper waters are more rarefied and less heavy than the vapors, why could they not be perpetually sustained by the fire together with air which are beneath, as the water, heavier, is sustained by only the air for a definite time? In fact it is clear that even the thick clouds and the heavy bodies of dragons and birds are held up by air. Nor do any of the believers doubt that human bodies, although of earthy nature, after the resurrection will be of such a lightness and thinness that they not only will be able to stand above the heavens but also to move without difficulties wherever the Spirit will have wanted. Moreover, who does not know that, if we put air in a bladder, this is able to raise and to support the skin, though it is lighter than the skin? Nay, the more the air is let in the greater is the mass it can raise. Analogously there is nothing to prevent that the jumble of air and fire, due to its lightness, be able to raise and sustain the waters in which it is included. Nor in any way should that diffuse water be able to flow downward, being surrounded on all sides by fire and air, since neither fire nor air offer any opening; because no body can take the place of another until the one has retreated from its own. And then all around the air and the fire are compressed by the surrounding waters so that they cannot pass away, and all around have upper waters, as in a sphere where all which is external is placed higher. In any case, so that air and fire can be confined and compressed, it is necessary that they be weighted by a moderate load such that they can sustain it. As is known, the heavy elements are only two, earth and water, and the earth is heavier than the water. Therefore, in order that the weight was more moderate, it was necessary to put the water on high rather than the earth, since the water could be more easily sustained by light bodies. Finally, who will be reasonably able to deny that in these bodies, be they animated or inanimate, which consist of the four elements, the elements merge in such a way that some particles of the lighter elements be arranged below

---

[46] Peter Abelard, *Sic et Non*, op. cit., Prologus, col. 1347 : *Ibi si quid veluti absurdum moverit, non licet dicere: Auctor hujus libri non tenuit veritatem; sed aut codex mendosus est, aut interpres erravit, aut tu non intellegis* (our Eng. trans.).

[47] É. Bréhier, *La Philosophie du Moyen Age*, op. cit., II, 2.

some particles of the heavier elements? In fact the lower parts of any great body are not quite devoid of fiery and airy particles, as, after all, neither the upper parts are immune to earth and water; but both the upper parts and the lower parts of this body consist of the four elements, all connected by such a moderate natural harmony, according to what is said with reference to God: '*You who connect the elements with the numbers*'. Why be astonished, then, if in the composition of the world he could superimpose on air and fire the water which is heavier than them and even obtain that it were sustained by them? Some have maintained that those upper waters are frozen and hardened like a crystal. And if things are thus, the more they are solid the more firmly they keep the fire and the air so that they still do not ascend, and the more they are sustained by them; on the contrary, perhaps it is not even necessary that they be sustained by them, since they are not fluid but solid, like a crystal. … Some maintain that these waters are retained for the Deluge, others, more correctly, say that they have been set in order to mitigate the fire of the stars. The blessed Augustine, leaving out all these opinions about the upper stars (that is, whether they are ice or not, or if they are of some avail), says: 'Certainly only the Creator knows of what type these waters are and to what employment they are reserved, what is not in doubt is their existence because of the witness of the Scripture'. Therefore, since so great a doctor remained in some doubt, an assertion on our part would appear very arrogant. Yet, since some said they were arranged and preserved for the inundation of the Deluge so they covered the earth falling abundantly, this is an absolutely foolish opinion.

"…What utility then the suspension of the waters can have, I think it is very difficult to decide since not even to the Holy Fathers it appeared established with a non-questionable reasoning. However, the opinion which appears to us to be the most probable (*probabilior…opinio*) is that according to which they have been constituted above all for mitigating the heat of the upper fire, so that its fieriness does not completely attract the clouds or the lower waters, the strength of the fire being such as to naturally attract humidity. … And it must be pointed out that where we say '*Let the firmament be in the middle of the waters*', the Hebrews have '*let there be some space under the waters*'; that is, an interval by which they are separated between themselves and thanks to which they never will come to touch".[48]

---

[48] Peter Abelard, *Expositio in Hexaemeron*, op. cit., cols. 741-744): *Fiat firmamentum in medio aquarum, et dividat aquas ab aquis. Firmamentum vocat aereum similiter simul et ethereum coelum, de quibus dictum est, et aquae quae super coelos sunt (Psal, CXLVIII, 4), quae utraque nunc interjacent inter has aquas interiores sive terram, et illas superiores aquas. Quod quidem ideo firmamentum dicitur, quod superiorem aquarum fluidam naturam ne inferius defluant et relabantur propria interpositione confirmat. Unde Hieronymus Samain Hebraice, id est coelos, ethereum scilicet atque aereum, quos hic firmamentum Moyses appellat, ex aquis dicit sortire vocabulum, id est pro eo sic appellatum, quod aquas ita superius firmat, ne inferius defluant. Quaeritur autem quomodo ignis et aer aquae substantiam quae ponderosior est sustentare valeant. Sed profecto tanta potest esse raritas atque subtilitas illarum aquarum, et tanta ignis et aeris massa quae ei subjacet, quod ab istis illae sustentari queant; sicut ligna et nonulli lapides ab aquis, quamvis ipsa terrenae sint et gravioris naturae. Quis etiam nesciat vicinum aerem, quamvis aquis levior sit, eas tamen exhalatione terrae vaporaliter tractas, antequam in guttis conglobentur, suspendere atque sustentare? Si ergo his aquis vaporatis, illae superiores rariores sunt ac minus corpulentae, cur non ab igne simul et aere subjacentibus sustentari perenniter valeant, sicut illae corpulentiores ab aere sol suspenduntur ad horam? Nam et nubes densas et ingentia draconum vel avium corpora ab aere sustentari manifestum est. Nec quisquam fidelium dubitat humana corpora, quamvis terrenae naturae sint, tantae subtilitatis ac levitatis, post resurrectionem futura, ut non solum super coelos consistere queant, verum etiam ubicumque voluerit Spiritus sine dilatione transferantur. Praeterea quis nesciat inclusum aerem in vescica circumstantem pellem vescicae undique suspendere atque sustentare, licet ipse omnino pelle illa sit levior? Imo etiam quantamcunque corporum molem sustentare posset quandiu inclusus cohiberi posset. Sic et illa aeris et ignis congeries, atque globus in illa aquarum corpulentia conclusus, nequaquam levitate sui eas suspendere vel sustentare impeditur. Nec ullo modo labi aqua illa circumfusa posset, qua undique ignem et aerem cohibet, donec ei ignis vel aer in aliquam partem cederet; quoniam locum unius corporis nullatenus alterum occupare potest, nisi illo primitus inter recedente. Undique vero aer ipse et ignis, ne devolare fortassis possint, circumstantibus aquis*

We have tried to report the essential excerpts of the commentary which Abelard devotes to the celestial waters, to show in details his thought and his way of solving the question. He expounds many of the interpretations and the explanations given up to then by the Fathers of the Church, and which we too have seen, insisting on relating the justifications of that situation based on the "lightness" of the waters. The only exponent explicitly quoted is, as one would expect, Augustine. But Abelard makes it clear that Augustine, so to speak, applies to the Creator for what concerns the nature of those waters and the use one should make of them ("*Sane quales ibi aquae sint, quosve ad usus reservatae, Conditor ipse noverit, esse tamen eas ibi Scriptura testante nulli dubium est*"). The final part of this phrase is what Abelard assumes as an undoubtable truth. In fact he has made clear in the Prologue of *Sic et Non* that what is maintained in the Scriptures cannot be questioned, the only thing one can do is to try to understand it, and Abelard does not feel like choosing an interpretation with certainty, since not even Augustine did it.

But he dares something more than Augustine with the adjective *probabilior*: the interpretation he opts for is not the correct one, but the more probable. Our comment, at this point, is that John Scotus (whose name is not even mentioned by Abelard) was more "creative".

2.2.2 Hugh of St. Victor (1096–1141)

In the twelfth century many schools flourished, mainly in France, and some of them had philosophers and theologians destined to characterize the history of the medieval thought. Besides the schools Abelard founded, in his intense activity as teacher and polemist, very famous were the School of Chartres and in Paris, at that time the capital of European culture, The Royal Abbey and School of Abbey of Saint-Victor, whose disciples and masters were named Victorines.

---

comprimuntur et undique superpositas habent aquas, quomodo in omni globo quae exteriora sunt superiora sunt, Ut autem cohiber illa et comprimere possint, aliquid gravitatis inesse necesse est, et ita moderate, ut ab illis queant sustentari. Constat vero, ut dictum est duo tantum elementa gravia esse, terram scilicet atque aquam et terra graviorem aqua. Unde ut moderatior esset gravitas aquam potius quam terram superponi oportuit, quae et facilius sustentari a levibus potuit. Enique in his corporibus, quae ex quatuor constant elementis, tam animatis quam inanimatis, quis rationabiliter negare possit, haec quatuor elementa tali sibi modo copulari, ut nonnullae particulae leviorum elementorum aliquibus particulis graviorum supponantur? Non enim alicujus magni corporis inferiores partes ita prorsus ignei vel aerii elementi admistione penitus carent, sicut nec superiores ab aqua et terra prorsus sunt immunes; sed tam superiores quam inferiores hujus corporis partes ex quatuor constant elementis, congrua quadam naturali moderatione sibi colligatis, juxta illud quod ad Deum dicitur: Qui numeris elementa ligas. Quid ergo mirum si in compositione mundi aqua igne vel aere gravior superponi potuit illis, et ab eis etiam sustentari? Nonnulli autem aquas illas superiores glaciali concretione soliditas atque in crystallum induruisse astruxerunt. Quod quidem si ita est, quanto magis sunt solidae, tanto vehementius conclusum ignem et aerem cohibent ne aliquo obscedant et tanto fortius ab ipsis sustentantur; imo fortassis nec jam ab eis sustentari eas necesse est, quae jam fluidae non sunt, sed in crystallum soliditate. ... Quidam ad inundationem diluvii reservatas, alii vero rectius ad ignem siderum temperandum suspensas affirmant: Beatus vero Aygustinus istas opiniones praetermittens de aquis illis superioribus, utrum videlicet glaciales sin vel non, vel quas in se habeant utilitates, ait; "Sane quales ibi aquae sint, quosve ad usus reservatae, Conditor ipse noverit, esse tamen eas ibi Scriptura testante nulli dubium est". Quod ergo tantus doctor quasi dubium sibi reliquit, affirmare nobis arrogantissimum videtur. Quod vero nonnulli opinati sunt eas ibi constitutas et reservatas ad inundationem diluvii, ut inde scilicet labentes abundantia sui terram cooperirent, frivolum omnino deprehenditur. ... Quid ergo suspensio illa aquarum utilitatis habeat, quod nec a sanctis certa sententia definitum est, difficillimum disseri arbitror. Illa tamen nobis probabilior videtur opinio, ut ob hoc maxime ad calorem temperandum superioris ignis constituerentur, ne fervor ille superior vel nubes ipsas vel aquas inferiores omnino attraheret, cum sit vis ignis naturaliter attractiva humoris. ... Et notandum quod ubi nos dicimus Fiat firmamentum in medio aquarum Hebraei habent: sit extensio infra aquas; hoc est intervallum quo ipsae ab invicem in perpetuum separentur ne se ulterius contingant (our Eng. trans.).

The most celebrated of the Victorines was Hugh of Saint Victor. One might say that Hugh was Abelard's younger contemporary, born almost twenty years later but dying a year before Abelard. He was born in 1096 at Hartingam in Saxony and educated at the cloister of Hamersleben close to Halberstad; he arrived at the monastery of Saint-Victor in 1115 and remained there until his death in 1141 (in 1139 he was raised to the rank of cardinal bishop of Frascati by Pope Innocent II). Among his several works (he was one of the main personalities of Augustinian scholastic philosophy), we shall refer to that which is considered his most important work: *De Sacramentis Christianae fidei*,[49] written in the last years of his life. It is the first great medieval theological *summa*. In it, inter alia, Hugh carried out a selection among the Catholic rituals, by choosing those which we nowadays call "sacraments" and which were assumed as dogmas by the Catholic Church in the Lateran Council IV of 1215.

Coming back to the subject we are interested in, Hugh devotes the first book of *De Sacramentis* to teaching how the Scripture must be read and discerning the *historical* reading from the *allegorical* reading: "Since, therefore, I previously composed a compendium on the initial instruction in Holy Scripture, which consists in their historical reading, I have prepared the present work for those who are to be introduced to the second stage of instruction, which is allegory. By this work, they may firmly establish their minds on that foundation, so to speak, of the knowledge of faith, so that such other things as may be added to the structure by reading or hearing may remain unshaken. For I have compressed this brief *summa*, as if it were, of all doctrine into one continuous work, that the mind may have something definite to which it may affix and conform its attention, lest it be carried away by various volumes of writings and a diversity of readings without order or direction".[50]

Let us see now how Hugh applies his method to the problem of the waters above the firmament. In the thirteenth chapter of the first part of the first book (*Quare aquae illae quae supra coelum sunt non dicit Scriptura quod sint congregatae in unum locum*) he writes, "Regarding those waters which are above heaven, Scripture has not said that they were gathered into one place, as in the case of those which were under the heaven. Great are the sacraments in all these matters, and not to be explained in the present summarized treatment: that the waters which are under heaven are gathered together into one place; that the dry land appears and brings forth plants, and that at the same time the very spaces of the air finally through the contraction of the dark mist are made clear, and that the courses of the waters are scattered everywhere for migrating; and moistening the earth through its body; and that nothing was done without cause.

"This seems strange, that the waters which are under heaven are gathered into one place, and that those which are above heaven are not gathered together and no place is assigned to them, but they are left diffused and spread out, as if the waters did not wish to be compressed or collected. What do you think this means, unless that 'the charity of God is poured forth in our hearts by the Holy Ghost who is given to us'? (Rom. 5.5) And these waters are above the

---

[49] Hugh of St. Victor, *De sacraments Christianae Fidei*, MPL 176, cols. 173-618 B. See the English translation: Hugh of St. Victor, *On the sacraments of the Christian faith*, trans. Roy J. Deferrari (Cambridge, MA, 1951; rpt. Wipf and Stock Publishers, 2007).

[50] Hugh of St. Victor, *Libri prioris De Sacramentis Prologus*, MPL 176, cols. 183-184: *Cum igitur de prima eruditione sacri eloqui quae in Historica constat lectione, compendiosum volumen prius dictassem, hoc nunc ad secundam eruditionem (quae in allegoria est) introducendis praeparavi; in quo, si fundamento quodam cognitionis fidei animum stabiliant, ut caetera quae vel legendo vel audiendo superaedificare potuerint, inconcussa permanent. Hanc enim quasi brevem quamdam summam omnium in unam seriem compegi, ut animus aliquid certum haberet, cui intentionem affigere et conformare valeret, ne per varia Scripturarum volumina et lectionum divortia sine ordine et directione raperetur* (Eng. trans. Roy J. Deferrari, op. cit., p. 3)

inhabitants of heaven, because, says the Apostle, 'I show unto you yet a more excellent way' (Cor. 12.31). 'If I speak with the tongues of men and of angels, if I should have all prophecy and should know all mysteries," what is this? (I Cor. parts of 1 and 2) it profiteth nothing, if have not charity' (I Cor. 13.2). 'The peace of God', he says, 'which surpasseth all understanding, keep your hearts and minds' (Philip. 4.7). Now in a certain way we see why those waters which are above heaven were not to be collected and compressed into one place, since charity ought always to be spread out and extended; and the more widely it is expanded, the more highly is it elevated. But the waters that are under heaven must be gathered together and constrained into one place, so that by fixed passages and definite outlets they may be conducted from there anywhere. For unless the lower affection of the soul is constrained by a definite law, dry land can not appear, nor can it produce plants, just as the Apostle chastises his body and brings it into subjection, lest perhaps, when he has preached to others, he himself may become a castaway (I Cor. 9.27)".[51]

It is clear that Hugh is distant from the interest in nature which, as we shall see, will be shown by the theologians of the School of Chartres. In this regard, in fact, he even contradicts himself by upholding the atomistic theory in *De Sacramentis*, whereas in the other his important work (*Didascalicon*, MPL 176, cols. 739-838), he maintains that the object of physics is the consideration of the four elements independent of their conditions of mixture.

Hugh favors an exclusively allegorical interpretation to the narration of the creation and interprets Genesis with the aid of the epistles of St. Paul. The supracelestial waters are identified with God's charity, again quoting St. Paul. In conclusion, it seems to us possible to say that Hugh, and perhaps his disciples as well, are an exception in the general tendency of that time.

2.2.3 Thierry of Chartres (1100?–1156?)

The most famous of the cathedral schools in the twelfth century was certainly the School of Chartres, whose first important representative was Bernard, master in the cathedral school between 1114 and 1119 and chancellor of the cloister between 1119 and 1124. None of his writings are extant and what biographical information about him we have comes from the work

---

[51] Hugh of St. Victor, *De Sacramentis Christianae Fidei*, op. cit., I, I, XXIII (MPL 176, col. 202): *De illis aquis quae supra coelum sunt non dixit Scriptura, quod congregatae sunt in locum unum; quemadmodum de illis quae erant sub coelo. Magna sunt in his omnibus sacramenta, nec explicabilia praesenti abbreviatione, quod aquae quae sub coelo sunt congregantur in unum locum, quod apparet arida et germina prpducit, simulque ipsa aeris spatia contracta tandem caligine serensntur, et aquarum tractus ad irrigandam atque infundendam terram per corpus illius quacumque sparguntur; et nihil sine causa factum est. Hoc mirum videtur, quod aquae quae sub coelo sunt congregantur in locum unum, et illae quae sunt supra coelum non congregantur neque eis locus unus tribuitur, sed relinquuntur diffusae atque expansae, quasi aquae coarctatari nolint vel colligi. Quid putas hoc sibi vult nisi quod* Charitas Dei *diffusa est in cordibus nostris per Spiritum sanctum qui datus est est nobis? (Rom. V). Et istae sunt super coelestes aquae; quia adhuc excellentiorem, inquit Apostolus, viam vobis demonstro (I Cor. XII).* Si lingua hominum loquar et angelorum; si habuero omnem prophetiam; si novero mysteria omnia, quid est? nihil prodest si charitatem non habeam *(I Cor. XIII).* Pax, ait, Dei quae exsuperat omnem sensum custodiat corda vestra et intelligentias vestras *(Philipp. IV). Jam quodammodo videmus quare aquae illae quae supra coelum sunt non debuerunt colligi tttet in unum coarctari, quoniam charitas amplificanda semper est et dilatanda; et quanto latius panditur tanto celsius sublimatur. Quam autem sub coelo sunt aquae, congregandae sunt et costringendae in locum unum, ut inde per certos tramites et exitus ordinatos, quacumque deducantur. Quoniam affectus animae inferior nisi certa lege costringatur, non potest apparere arida, nec germina producere, sicut Apostolus. Castigat corpus suum et in servi utem religit, ne forte cum aliis predicaverit, ipse reprobus fiat* (I Cor. IX) (Eng. trans. Roy J. Deferrari, op. cit., pp. 24-25).

of John of Salisbury (1120–1180), who calls him "the most perfect among the Platonists of our century.[52]

As we know, the work of Plato known and studied at that time was the *Timaeus*, in the commentary and the partial translation by Calcidius. In the philosophy of the School of Chartres, the "mixing" of Platonism with Genesis, which had already occurred in the preceding centuries, continues and an interest in nature is expressed, in clear contrast with the mysticism of the Victorines. This interest also induces one to promote a *historical* reading of Genesis.

We can now see an example of this in the work of Thierry (or Theodoricus), Bernard's younger brother. Both both dates of Thierry's birth and death are uncertain. According to Abelard, he was already *magister* at Chartres in 1121. It is also probable that he lived in Paris between 1124 and 1141 and, when he was elderly, he retired to a monastery. Among the most important of his works is the *Tractatus de sex dierum operibus*,[53] which he begins by saying that he will explain the first part of Genesis following natural science (*secundum physicam*) and neglecting the allegorical interpretation already given with clarity by the Holy Fathers. In effect, he tries to rationally explain the formation of the world through the reading and interpretation of the first chapters of Genesis.

Let us see how he deals with the problem of the waters above the firmament, in the narration of the second day of the creation: "And once the air was illuminated by the power of the higher element [fire], it followed naturally that, by illuminating the air, fire heated the third element, water, and by heating it suspended it as a vapor above the air. For it is the nature of heat to divide water into very tiny drops and raise those tiny drops above the air by the power of its motion, as is seen in steam in a boiling pot and as appears in the clouds in the sky. For the clouds or steam are nothing other than masses of very tiny drops of water raised into the air by the power of heat. If the power of heat becomes stronger, the whole mass changes into pure air. If, however, it becomes weaker, then those very tiny drops, rushing together, make bigger drops and cause rain. But if those tiny drops are compressed by wind, they cause snow, and if they become large, they make hail.

"Therefore the huge mass of moving water, which in the beginning doubtless reached to the region of moon, was suspended by heat above the height of the ether, so that immediately, during the second rotation of the heaven, it happened that the second element, air, was in the middle between the moving water and the water suspended as a vapor. This is what the author refers to when he says, *And he placed the firmament in the middle of the waters* (Genesis 1. 6). Then it was fitting to call the air the 'firmament', since it firmly supported the water above it and restrained the water below it, making sure that they did not cross over into one another. Or perhaps air is called the 'firmament' rather because by its lightness it firmly restrains the earth from every direction and forms it into a hard ball. For there is a reciprocal relationship between the hardness of the earth and the lightness of the air: The hardness of the earth comes from the constriction of the light air, and the lightness and mobility of the air has its substance from the fact that it rests on the stability of the earth".[54]

---

[52] John of Salisbury, *Metalogicus* (MPL 199, cols. 823-946C), IV, 35C (col. 938): *Bernardus quoque Carnotensis, perfectissimus inter Platonicos saeculi nostri* (our Eng. trans.).

[53] The Latin text of this work (in a critical edition) is published in *Commentaries on Boethius by Thierry of Chartres and his school*, ed. Nickolaus M. Haring (Toronto: Pont. Inst. of Medieval Studies, 1971).

[54] Thierry of Chartres, *Tractatus de sex dierum operibus*, 7-8: *Aer uero ex superioris elementi uirtute illuminato, consequebatur naturaliter ut, ipsius aeris illuminatione mediante, confaceret ignis tercium elementum i. e. aquam et calefaciendo suspenderet uaporaliter super aera. Est enim natura caloris aquam in minutissimas guttas diuidere et eas minutas uirtute sui motus super aera eleuare sicut in fumo caldarii apparet: sicut etiam in nubibus celi manifestum est. Nubes enim siue fumus nichil est aliud quam guttarum aque minutissimarum congeries per*

In the beginning of the work, before the excerpt we have reported explaining how, on the second day, the waters above firmament were formed, Thierry says that Moses explains in a rational way the causes by which the world and the order of times of its institution originated. Obviously, it is not Moses but Thierry who sets up a reading *secundum physicam* of the text of Genesis. He entrusts the heat and the rotation of the heavens with the formation of the cosmic order and the birth of the living. The rotation of the heavens also scanned the days. In fact he says that the first rotation of the fire lightened the air and the duration of this lighting was "the first day". In the same way, the second rotation of the same fire, through the air, warmed the water and placed a firmament between water and water: the duration of this rotation was called "the second day".

The excerpt we have reported gives us a narration of what happened on that second day. The impression one draws from it is that Thierry constructs an interpretation—or better, an explanation—which aims to look "tenable", in accordance with common sense of that time. In fact, he appeals to examples familiar to all because they are drawn from everyday life, such as the water boiling in a pot. In later chapters of this short treatise, Thierry returns to the beginning of Genesis and formulates more "philosophical" questions, but this goes beyond our scope.

2.2.4 William of Conches (1080 ca.–1145 ca.)

Another important representative of the School of Chartres was William of Conches. Born around the end of the eleventh century in a little town of Normandy, he was a pupil of Bernard of Chartres and later the teacher of John of Salisbury. In all probability, he taught at Chartres for his whole life, although some have supposed his presence at Paris for a certain period. William is considered a typical representative of the so-called "Chartrian Platonism". His principal works are *De philosophia mundi*,[55] an extensive philosophical and scientific encyclopedia, and *Dragmaticon philosophiae*,[56] a dialogue about natural philosophy between a philosopher (William's alter ego) and the Duke of Normandy.

In the prologue of the *Dragmaticon*, he points out that he will repudiate part of the statements made in *De philosophia mundi*. As we shall see, this did not happen for his opinion about the waters above the firmament; his conviction regarding their non-existence as congealed waters remains unchanged in both works. In the first work the opinion of William is quite resolute and seems to be addressed to generic interlocutors ("Some say that..."), where, as will become

---

uirtutem caloris in aera eleuata.. Sed si uirtus caloris uehementior fuerit, tota illa congerries in purum aera transit: si autem debilior tunc mirum gutte ille minutissime semet inuicem incurrentes grossiores guttas faciunt: et inde pluuia. Quod si minute ille gutte uento constricte fuerint inde nix: si uero grosse inde grando. Magnitudo igitur aquarum labilium que nimirunm usque ad regionem lune in principio ascendebat ita per calorem super summum etheris suspensa est ut statim in secunda celi conuersione ita contingeret quod secundum elementum i. e. aer esset medium inter aquam labilem et aquam uaporaliter suspensam. Et hoc est quod dicit auctor: et posuit, "firmamentum in medio aquarum". Et tunc aer aptus fuit ut, "firmamentum"appellaretur quasi firme sustinens superiorem aquam et inferiorem continens: utramque ab altera intransgressibiliter determinans. Uel potius, "firmamentum" dicitur aer eo quod terram leuitate sua ex omni parte firme coherceat et in hanc duriciam conglobet. Est enim ista reciprocatio inter terre duriciem et aeris leuitatem ut durities terre ex circumstrictione leuis aeris proueniat: leuitas uero aeris atque mobilitas ex eo quod terre stabilitati innitur habet substantiam (Eng. trans. Katharine Park, available at https://www.academia.edu/31388090/Thierry_of_Chartres-Treatise_Six_Days-trans._Park.pdf, accessed 12.01.2020).

[55] William of Conches, *De philosophia mundi libri quatuor*, MPL 172, cols. 39-102A (the MPL attributes it to Honorius of Autun).

[56] William of Conches, *Dragmaticon philosophiae*, in *Guillelmi de Conchis Opera omnia*, vol. I, ed. I. Ronca (Brepols, 1997).

evident in the second work, the target is Bede, who maintained the argument of the congealed waters in the *Hexaemeron* (see the excerpt we have quoted in 2.1.2).

Let us see what he says in the *De philosophia mundi*. He begins with, "Some say that over the ether there are congealed waters, taut like a skin, which appear to our eyes, and over them waters whose existence is confirmed by the authority of the Holy Scripture, which says: *He put the firmament in the midst of the waters,* and again, *He divided the waters which are under the firmament from those which are over the firmament* (Gen. 1). But, since this is against reason, we show why things cannot be in this way and then how the aforesaid sentences of the Holy Scripture must be understood. If in that place there are congealed waters, then there is something weighty in itself, but the natural place of the weighty things is the earth. Moreover, if in that place there are congealed waters will they be joined to the fire or not? If they are joined to the fire, it being warm and dry, and the congealed water cold and humid, the opposites would be joined without any intermediary; then there would never be concordance but the repugnance of the opposites. Even more: if the congealed water is joined with the fire either it will be dissolved by the fire or it will extinguish the fire. So, if the fire and the firmament remain, where are the congealed waters, joined with the fire; if they are not joined, is there something between them? But what? An element? But no element made by elements and then visible, can stay over the fire. Thus, not being there anything visible, it follows that there are no congealed waters".[57]

He reaffirms his opinion with: "I already know what they will say: we don't know how this can happen, but we know that God can let it happen. Stupid! Can one say something more stupid than this? Since God can do this, we should not see if the thing has been made, nor try to find the reason of its being made, nor show its usefulness. In fact, God does not make all that he could make. By way of a rustic illustration, God can let a calf to be born from a trunk: when has he ever done such a thing? Therefore, let them show the reason or the usefulness of what they are speaking or else let them give up maintaining that things are like that, since congealed waters do not exist there, nor other waters above them. When the Holy Scripture says: *He divided the waters which are under the firmament from those which are above the firmament* (Gen. 1.7), firmament is called the air, which consolidates and tempers the earth thing. Above this air there are waters hanging as clouds of vapors, as we shall show later, which are divided from those which are under the air. Analogously one must interpret: *He put the firmament in the midst of the waters,* although we think that this has been said more allegorically than *ad litteram*".[58]

---

[57] William of Conches, *De philosophia mundi*, II, II, col. 57: *Dicunt enim quidam super aethera esse aquas congelatas, quae in modum pellis extensae occurrunt oculis nostris, super quas aquae sun confirmatae, hac autoritate divinae paginae, quae ait:* Posuit firmamentum in medio aquarum *et iterum:* Divisit aquas, quae sunt sub firmamento ab his quae sunt super super firmamentum (Gen. 1). *Sed, quoniam istud contra rationem est, quare sic esse non possit ostendamus, et qualiter divina Scriptura in supradictis intelligenda sit. Si ibi sunt aquae congelatae, ergo aliquid ponderosum et grave? Sed primus locus ponderosorum et gravium est terra. Si iterum ibi sunt aquae congelatae, vel igni conjunctae, vel non? si igni cojunctae sunt, cum ignis calidus et siccus sit, aqua congelata et humida, contrarium sine medio suo contrario conjunctum est: numquam ergo ibi concordia, sed contrariorum repugnantia? Amplius: Si aqua congelata conjuncta est igni vel dissolvetur ab igne, vel extinguit ignem. cum ergo ignis et firmamentum remanent, ubi sunt aquae congelatae, conjunctae igni, si conjunctae non sunt, aliquid inter eas et ignem est? Sed qid? elementum, sed nullum superius igne factum est ex elementis, visibile, ergo. Unde igitur non videtur. Restat ibi non esse aquas congelatas* (our Eng. trans.).

[58] William of Conches, *De philosophia mundi*, col. 58: *Sed scio qui dicent. Nos nescimus qualiter hoc sit, scimus Deus posse facere. Miseri! Quid miserius quam dicere istud est? quia Deus illud facere potest, nec videre sic esse, nec rationem habere quare sic sit, nec utilitatem ostendere ad quam hoc sit. Non enim quidquid potest Deus facere, hoc facit. Ut autem verbis rustici utar, potest Deus facere de trunco vitulum: fecitne unquam? Vel igitur ostendant rationem, vel utilitatem ad quam hoc sit, vel sic esse indicare desinant. Si ergo aquae ibi congelatae non sunt, nec super eas aliae. Cum autem divina pagina dicat:* Divisit aquas, quae sunt sub firmamento, ab his quae sunt super

In the *Dragmaticon*, written when he was at a ripe age, he will return to the subject, making his interlocutor (the Duke) maintain Bede's theory:

Duke: Bede says that what appears to us like a taut skin are congealed waters.

Philosopher: In those things which concern the Catholic faith or moral education it is not permissible to contradict Bede or any other of the Holy Fathers. However, in those things which concern physics, if they are wrong in something, it is permissible not to agree. In fact, even though greater than us, they were nevertheless men.

Duke: When an inferior contradicts one who is superior to him, he is obliged to indicate the reason why the things must be otherwise. If, then, you want to be believed in this regard let the reason, true or verisimilar, be shown by which the things cannot be as Bede says.

Philosopher: The water by its nature is weighty; if congealed, since it is in this case closer to the nature of the earth, it would be even weightier and the motion peculiar of what is weighty is toward the center. Thus, if there were congealed waters over the ether, because of natural gravity they would descend down to the deepest place.

Duke: We see that the heavy bodies are supported by the vault of the arch and never fall: thereby these congealed waters, being supported as though by a vault cannot fall.

Philosopher: If the arch which supports the vault had no basements leaning on something solid, it would inevitably fall at once. Hence, this arch you imagine rests its basement on something solid, but there is nothing except the earth. Therefore, it is necessary that those basements are fixed to the earth: and this is worse than a stupid shaft of wit.[59]

Therefore, the two representatives of the School of Chartres we have considered favor a *physical* solution and, in the case in point, consider the "waters above the firmament" as clouds of water vapor poised on the air, like the clouds, which as all can see, in the earth's atmosphere. If one holds the text of Genesis to be an absolute and indisputable truth, then this solution appears as the most "plausible" reading *ad litteram* according to the physical knowledge of that time.

---

firmamento *(Gen. 1,7), aera firmamentum vocavit, quia firmat et temperat terrena. Super hanc aquam sunt, vaporaliter in nubibus suspensae, ut in sequentibus ostendetur, quae divisae sunt ab his quae sunt sub aere. Similiter exponatur: Posuit firmamentum in medio aquarum, quamvis hoc plus allegorice, quam ad litteram dictum credimus* (our Eng. trans.).

[59] *Dragmaticon*, III, 2: *DUX: Beda dicit aquas congelatas esse quae ad modum pellis extensae nobis apparent. PHILOSOPHUS: In eis quae ad fidem catholicam uel ad institutionem morum pertinent, non est fas Bedae uel alicui alii sanctorum patrum contradicere. In eis tamen quae ad physicam pertinent, si in aliquo errant, licet diuersum adfirmare. Etsi enim maiores nobis, homines tamen fuere. DUX: Quotiens minor maiori contradicit, rationem, quare aliter esse oporteat, oportet inducat. Si uis igitur ut super hoc tibi credatur, quare sic esse non possit ut Beda dicit, uel uera uel uerisimilis ratio adducatur. PHILOSOPHUS: Omnis aqua naturaliter est ponderosa; congelata, quia plus accedit ad naturam terrae, est ponderosior, motus proprius ponderum est ad centrum. Si igitur aquae congelatae super aethera essent, naturali gravitatae ad ima descenderent. DUX: Videmus grauia per fornicem arcus sursum teneri neque umquam descendere: unde aquae istae congelatae, quia in modum fornicis sunt suspensae, descendere non possunt. PHILOSOPHUS: Arcus fornicis, nisi bases quae alicui solido innitantur habeat, necesse est statim corruat. Iste igitur tuus arcus, quem confingis, alicui solido basem figit, sed nichil est praeter terram solidum. Oportet igitur quod istae bases terrae inhaerant: quod utilitatem scurrilis ioci excedit* (our Eng. trans.).

2.2.5 The Summary of the Pseudo-Bede

Also belonging to the twelfth century is a sort of text, perhaps datable according to the scholars around 1140, bearing the title *De Mundi coelestis terrestrisque constitutione liber*,[60] inserted in Migne's *Patrologia Latina* among the works of Bede the Venerable but considered *dubia et spuria*. It is a concise exposition of the subjects dealt with in Bede's *De Natura rerum* (perhaps for this it was attributed to him, albeit without certainty); the unknown author is referred to as "Pseudo-Bede". The text is undoubtedly associable with the *physical* production of the School of Chartres..

Let us see how it summarizes the subject we are interested in: "The supracelestial waters. The first assumption is that in the external surface of the heaven, as in that of the earth, there are humps and hollows in which the waters are said to be contained; and they rotate with such a velocity that it prevents them from falling down, as can be shown with any filled vase. In fact, the greater the force with which [the vase] is revolved by the hand, the smaller the quantity [of water] which spills from it. The second assumption is that the waters are contained [above the firmament] in the form of vapors, as the clouds above us. The third assumption is that, due to the remoteness of the sun, from which the heat mainly comes, they cohere to the heaven made hard by the cold. Thus they say that Saturn is the coldest of the planets, being closest to those waters. There is also a fourth assumption, according to which the firmament does not turn since it is fixed, but it is the stars, placed on it and not fixed, which turn. And one wonders if it is round or square. At last, the problem can be solved by the omnipotence of God, because [those waters] would be kept there following God's will, even if how is far from being clear to men, and preserved by God for the use of divine beings, or for moderating the fire of the heaven. Above these [waters] are the spiritual heavens where the angelic virtues are encapsulated. According to others there is nothing, if not the vacuum".[61]

As it is easy to see, the author favors a *physical* reading of the two verses and suggests to us all the possible explanations that he is aware of. The recourse to centrifugal force and the related example is interesting.

2.2.6 The Encyclopedists of the thirteenth century

The thirteenth century is that in which the translations from the Greek and from the Arab of the works of the Greek philosophy and science are disseminated throughout Europe. The activity of the translators, who were from different countries, started in Spain around the end of the previous century and the translations from Aristotle's works dominated the scene and prompted commentaries and new discussions.[62] These new "acquisitions" also brought about

---

[60] Bede (attributed), *De Mundi coelestis terrestrisque constitutione liber*, MPL 90, cols. 881-910A.

[61] Ibidem, col. 893: *De supercoelestibus aquis. Prima conjectura est, quod sint in exteriori superficie coeli extantia et subsidentia loca, ut etiam in superficie terrae, in quibus aquas contineri dicunt; cum tanta vero celeritate volvuntur, ut non labantur, quod de pleno vase quolibet probari potest. Nam quo maiori impetu manu circumvolvitur, eo minus effunditur. Secunda est, quod ibi vaporaliter quemadmodum hic nubes pendent, contineantur. Tertia est, quod propter solis remotionem, a quo praecipuus calor exhibetur, indurato glacialiter coelo cohaereat. Unde dicunt Saturnum frigidissimum planetarum, quia illis aquis vicinissimum. Est etiam quarta, firmamentum ut fixum non volvi, sed stellas illi appositas, nec fixas moveri; et hi an quadratum, an rotundum sit dubitant. Novissime quaestio divina virtute solvitur, quia ibi teneantur secundum Dei voluntatem, licet hic modus homines lateat, quas ibi in usum divinorum servavit Deus, vel ut ignem coeli temperent: super has spirituales coeli sunt, in quibus angelicae virtutes continentur. Secundum alios nihil, nisi inane* (our English trans.).

[62] On this subject, there are works almost classical such as those of H. Haskins, *Studies in the History of medieval Science* (2nd ed. Harvard University Press,1927) and A. C. Crombie, *Augustine to Galileo: The History of Science A.D. 400–1650* (Heinemann, 1952; rpt. as *The History of Science from Augustine to Galileo*, Dover, 1959).

the renewal and "modernization" of the whole corpus of notions handed down through the encyclopedic compilations. The last important work in this field, the *Etymologiae* of Isidore of Seville, dated back to the beginning of the seventh century and, obviously, at a distance of more than five hundred years, showed all its inadequacy.

Starting from the beginning of the thirteenth century, new authors appeared. Their encyclopedias had a structure different from that of the encyclopedias of the past and, in part, new purposes as well. In recent decades of our own day, scholars have turned their attention to the medieval encyclopedias (quite neglected in the past) with renewed interest.[63] There are five encyclopedists who are generally considered the most important: Alexander Neckam, Bartholomew of England, Arnoldus of Saxo, Thomas of Cantimpré and Vincent of Beauvais.

Like the other personages we have already come across, these also belonged to religious orders and their works, besides to be used to educate the brothers, were also aimed at supplying subjects and explanations for the sermons. In these works, the subject of the creation continued to be of fundamental importance and, in certain cases, the structure of the work was derived from the Hexaemerons of the past. Roughly proceeding in chronological order, in these works, as usual, we shall investigate if and how the subject of the waters above the firmament is dealt with.

Alexander Neckam (1157–1217)
On the childhood of Alexander Neckam one usually relates a bizarre anecdote which we too shall report, Thorndike's narration: "In the year 1157 an English woman was nursing two babies. One was a foster child; the other, her own son. During the next fifty years these two boys were to become prominent in different fields. The fame of the one was to be unsurpassed on the battlefield and in the world of popular music and poetry. He was to become king of England, lord of half of France, foremost of knights and crusaders, and the idol of the troubadours. He was Richard, Coeur de Lion. The other, in different fields and a humbler fashion, was none the less also to attain prominence; he was to be clerk and monk instead of king and crusader, and to win fame in the domain of Latin learning than Provençal literature. This was Alexander Neckam".[64]

He was born at St. Albans (Hertfordshire), finished his first studies in the abbey of St. Albans and then taught at Dunstable. He later moved to Paris, at the beginning as a disciple and then as a lecturer at the university, where he became a master of dialectics. Afterwards he went back to England and was master at Dunstable and lecturer at the University of Oxford. In 1213 he became canon of Circencester. He died at Kempsey (Worcestershire) in 1217. He wrote works on all the branches of medieval knowledge. We shall only deal with one of his works in particular, *De Naturis rerum* (*The Natures of Things*), which consists of five books in all, the first two of which form a separate treatise.[65] Regarding this work Lynn Thorndike says: "The brevity of *The Natures of Things*, which consists of but two books, if we omit the other three of its five books which consist of commentaries upon Genesis and Ecclesiastes, hardly allows us to call it an encyclopedia; but its title and arrangement by topics and chapters closely resemble the later works which are usually spoken as medieval encyclopedias."[66] In fact, as we have

[63] See, for instance, the part regarding The Medieval Latin Encyclopedias in the volume *The Medieval Hebrew Encyclopedias of Science and Philosophy. Proceedings of the Bar-Ilan University Conference*, ed. Steven Harvey (Springer, 2000).

[64] L. Thorndike, *History of Magic and Experimental Science*, op. cit., p. 188.

[65] The Latin text of the first two books can be found in Alexander Neckam, *De Naturis rerum libri duo: with the poem of the same author, De laudibus divinae sapientiae*, ed. Thomas Wright (London, 1863).

[66] L. Thorndike, *History of Magic and Experimental Science* op.cit.,vol. II, p. 193.

recalled above, it is usually inserted into the category of thirteenth-century encyclopedias (even though it was actually written around the end of the twelfth).

In the first book, after four chapters dealing with the creation, Alexander devotes twelve chapters to subjects of "astronomical" nature and consequently also the waters above the firmament are dealt with. However, before reporting the excerpt where this is done, we wish to point out a few things.

The most important is that the then recent translations of Aristotle's works (but also of other authors, Ptolemy included) had completely changed the intellectual context in which the monks like Alexander moved. The *De Naturis rerum* at several times relates Aristotle's thought: *De caelo et mundo* and other works are quoted; Platonism, always present in the works of the earlier authors, is no longer prevailing. Alexander, as many others before him, introduces the Holy Trinity into the narration of the creation and, to that end, studies Hebraic in order extract arguments in favor of his reading from the ancient Hebraic version of the Bible.

This appears to be a choice for an allegorical reading but, oddly enough, when speaking about a divergence between Aristotle's theory and the Bible regarding the size of the planets, he sets out an opinion which may appear "modern", or even "laic": "To what we said above about the size of the planets, it seems contrary to what one reads in Genesis: 'God made two great lights: one greater for lightening the day, and one smaller for governing the night'. Indeed, according to the series of those mentioned above, it does not appear that the moon must be included among the great stars. But historical narration [*historialis narratio*, i.e., in Genesis] follows the impression of vision and common opinion, though Moses did not speak about greatest lights, but about great lights".[67]

If we are not stretching the text too much, one might interpret it as: Moses could not describe the things as they were (that is, according to Aristotle!) since the reader would not have understood them, because he wished to interpret what he was seeing following common sense. In other words, Moses does not express himself through allegories, but through a picture of the world which can be understood by an ordinary person of faith who reads the Bible. But we grant that our interpretation may be too far-fetched.

A little further on in the text, Alexander, speaking about the voice, says, "Even though I do not think that the voice is air, nonetheless without the action of the air it could be neither uttered nor heard".[68] If we read this sentence with a modern sensibility, the author is saying us that sound propagates only through a material medium.

Coming back now to the waters above the firmament, we find that in Chap. XLIX of the second book bearing the title *Quod aqua non sit inferior terra*, Alexander says: "One will object that the prophet says: 'God has consolidated the earth over the waters'. From this it would seem that one can think that the waters are below the earth, whereas Alfraganus says that the sphere of the waters and the earth is only one. The interpreters of the Holy Scripture attribute the phrase of the prophet to the use of everyday language, as it is usual to say that Paris is founded on the Seine. But the truth is that Eden is above the waters and also above the lunar sphere. Consequently, even the waters of the Deluge did not cause any trouble for Paradise. Enoch, who was already in Paradise even then, did not notice any growth of the waters because of the

---

[67] Alexander Neckam, *De Naturis rerum*, op. cit., I, XIII, pp. 49-50: *Eis vero quae de quantitatibus planetarum supradiximus videtur esse contrarium quod legitur in Genesi, "Fecit duo magna luminaria, luminare majus ut praeesset diei, et luminare minus ut praeesset nocti". Luna enim inter magna luminaria, secundum sopradictorum seriem non videtur annumeranda. Sed visus judicium et vulgarem opinionem sequitur historialis narratio, quamquam non maxima luminaria dixerit Moyses ista duo, sed magna* (our Eng. trans.).

[68] Alexander Neckam, *De Naturis rerum*, op. cit, p. 65: *Etsi enim vocem non credam esse aerem, tamen sine aeris beneficio nec proferri potest nec audiri* (our Eng. trans.).

Deluge. Indeed, the sea is at a higher level than the seashores, as vision shows. On that account, one must attribute the fact that the sea does not overstep the limit of the divine will fixed by God. ... Let no one misinterpret what I have said at the beginning and believe that I want to teach that the water is not below the earth, by using a language proper to the common people. Is it not actually true that one says that the Antipodes are under our feet? But if you want to speak philosophically, they are under our feet neither more nor less than we are under their feet. But did the Antipodes come perhaps from the remote progenitors? Whereas according to Augustine there are no Antipodes, but usually they are mentioned in theory or as a fiction".[69]

We recall that already in the fifth chapter of the first book (*De firmamento*), Alexander had established a parallelism between the firmament and the Church and consequently between the waters above the firmament and the baptismal waters. He returns to this parallelism in this chapter as well. In fact, in that part we have not reported-p. 159) he says: "Holy land for the Church was founded on baptismal waters...".[70] For the rest, it still seems to us it is not possible to speak *stricto sensu* of an allegorical interpretation, even though we are far from a reading *secundum physicam*, in spite of the appeal to the vision (*ut visus docet*). To conclude, we think that Alexander, rather than expanding on the question of the incongruity of the physical disposition of the elements, only manages to suggest possible explanations.

Bartholomew of England (1203?–1272)

We know very little about the life of Bartholomew of England (Bartholomaeus Anglicus). Undoubtedly he was an Englishman, as demonstrated by the appellative *Anglicus* that always appears in the manuscripts and by cross-checks of the quotations in his work and in the works of his contemporaries. As regards his birth date, one can only establish that it must be prior to 1203. He was a member of the Franciscan order, studied at Paris and moved to Magdeburg (Saxony) in 1231. It seems that this was where he wrote, around 1240, the work for which he is well known: *De genuinis rerum coelestium, terrestrium et inferarum proprietatibus* (usually quoted as *De proprietatibus rerum*). It is a work of nineteen books, the first three of which regard philosophy and theology (on God, the angels, the soul) and the other sixteen of which regard the natural sciences.[71] The eighth book (*De proprietatibus mundi et corporibus coelestibus*) is the one we are interested in for our purpose. In it the sentences of the Bible are fitted in with Aristotle's theories and the astronomical knowledge of that time. Let us see the third chapter (*De caelo aquaeo sive crystallino*): "The sixth heaven is aqueous, or crystalline,

---

[69] Alexander Neckam, *De naturis rerum*, op. cit., II, XLIX: "*Movebitur aliquis super hoc quod dicit propheta, 'Dominum firmasse terram super aquas'. Ex hoc enim videbitur haberi posse aquas esse inferiores terra, cum tamen Alfraganus dicat, unam esse spheram aquarum et terrae. Sancti igitur expositores referunt illud prophetae ad cotidianum usum loquendi quo dici solet Parisius fundatam esse super Secanam. Rei tamen veritas est quod paradisus terrestris superior est aquis, cum etiam lunari globo superior sit. Unde et aquae cataclysmi paradiso nullam intulere molestiam. Enoc, qui in paradiso jam tunc erat collocatus aquarum non sensit diluvii incrementa. Mare vero superior est litoribus, ut visus docet. Unde divinae jussioni attribuendum est, quod metas positas a Domino non trasgreditur mare. ... Quod vero in rubrica dixi nullus perperam intelligat, credens me docere velle aquam non esse sub terra eo loquendi genere quo vulgus uti solet. Nonne enim et antipodes sub pedibus nostris esse dicuntur. Si tamen philosophice loqui volueris, non magis sunt sub pedibus nostris qum nos sub pedibus eorum. Sed numquid de primis parentibus descenderunt antipodes? Secundum Augustinum non sunt antipodes, sed doctrinae causa aut figmenti ita dici solet* (our Eng. trans.).

[70] Alexander Neckam, *De Naturis rerum*, op. cit, p. 159: *Terra namque sanctae ecclesiae fundata est super aquas baptismales...* (our Eng. trans.).

[71] For the Latin text of this work, since a complete critical edition is still underway, one usually refers to the seventeenth-century edition *Bartholomaei Anglici De Genuinis Rerum Coelestium, Terrestrium et inferarum Proprietatibus* (Frankfurt: Wolfangum Richterum, 1601; rpt. Frankfurt: Minerva, 1964).

having been formed by the waters placed over the firmament by the divine power; in fact the authority of the Holy Scripture hands down to us that the waters were placed over the heavens and thus transformed and on this account they stay fixed there. However Bede says that those celestial waters, by divine nature, are hanging over the firmament not as tenuous vapor but because of a certain solidity, and thus to moderate the strength of the firmament, or to contain the heat produced by its whirlwind movement. In fact Bede held, contrary to the Platonists, that the heaven is of igneous nature. Whence Bede says: 'the heaven is thin and of igneous nature, round and equidistant from the centre of the earth'. And perhaps for this reason Bede thought that waters were necessary there for slowing down that celestial heat, so that the lower world could not be dissipated by such an inflammation. In fact, some say that the coldness of the star of Saturn derives from the natural coldness of those waters placed over the summit of the firmament, because of its closeness to it. They also say that the firmament, cooled by virtue of those waters, cools the orb of Saturn which is closer. But how this can happen rationally is not clearly manifest to those who reason. In fact, since the substance of water by reason of both its qualities, that is, the humidity and the coldness, is almost fully adverse to the celestial substance, it is far from clear to philosophers how among such different bodies it is possible to achieve a certain kind of unity or concordance. However it is written: 'He who generates concordance in his highest regions' (Job. 12). Hence the moderns think in another way, by analyzing more deeply, as I think, the interior matter of philosophy. In fact Alexander says that those waters which are over the heavens are not placed there as cold, fluid and humid, or rather as solid, frozen and weighty. In fact these properties are also contradictory among themselves and mutually opposed, but those waters are placed over the firmament under the order of the divine wisdom on the strength of their noblest nature, since they are especially close to the celestial nature. And these undoubtedly are the properties of the [celestial] nature: limpidity and transparency, which are principally and substantially in the nature of water which, for this reason, has affinity both with the empyrean heaven and the firmament. Whence God placed the water low on account of the coldness and the humidity and the other conditions necessary to generation and corruption, but made them transparent, since this was necessary to the conservation of all things. Therefore, he says that the heaven is called aqueous and crystalline by reason of mobility and transparency. In fact, it is transparent as crystal, since it receives from the superior heaven, that is from the empyrean, the light or the plenitude of the brightness and diffuses it in the lower things. Therefore, the heaven, almost invisible and hidden to us, is called crystalline not because it is solid like a crystal, but because it is uniformly bright and transparent. Instead, one calls aqueous what moves in the same way as water due to its thinness and mobility, and that motion transmits the movement to the near heaven and this in turn moves what is nearest to itself. And thus, the heaven which moves the lower things is chiefly protective of their motion, as Alexander says".[72]

---

[72] Bartholomew of England, *De Genuinis Rerum Coelestium, Terrestrium et inferarum Proprietatibus*, op. cit., VIII, III, pp. 378-379: *Sextum coelum est aqueum sive crystallinum, quod ex aquis positis super firmamentum divinitatis potentis est formatum, aquas enim esse super coelos collocatas divinae scripturae autoritas nobis tradit, quae ita sunt levigatae et subtiliatae, qod in materiam coelestem sunt conversae, et ideo permanent ibi fixae. Beda tamen docet quod aquae illae coelestes non vaporali tenuitate, sed gracili quadam soliditate, virtute divina super firmamentum sunt suspensae. Et hoc modo ad impetus firmamenti moderationem, vel ad caloris generati ex eius velocissimo motu repressionem. Opinio non fuit Bedae quod coelum igneae sit natura, sicut Platonici posuerunt. Unde dicit Bedae. Coelum est subtilis et igneae naturae, rotundum a centro terrae, aequalibus spaciis collocatum. Et ideo forte Bedae videbatur quod ideo fuit necesse ibi esse aquas, ut calor ille coelestis ad temperantia duceretur, quod ex tali inflammatione mundus inferior dispendium non pateretur, ex frigiditate enim naturali illarum aquarum super firmamentum verticem positarum, dicunt aliqui stellam Saturni esse frigidam propter illam, quam habet ratione situs cum firmamento vicinitatem. Dicunt etiam quod firmamentum per virtutem illarum aquarum*

Considering that the Alexander quoted by Bartholomew is Alexander of Hales (1185–1245), also a Franciscan, among the first who tried to reconcile Christian theology with the new translations of the works of Aristotle and the Arabs, let us try to understand what Bartholomew's opinion is. At present, scholars are discussing whether he is an original thinker or a compiler, that is, whether he introduces personal ideas in his work or he limits himself to juxtaposing what he draws from the ancient works and from the recent translations of Aristotle's works and from the Arabs.[73] It seems to us that, in this case, Bartholomew has not made any attempt to introduce new elements, except for a shy expression of a doubt: *"But how this can happen rationally is not clearly manifest to those who reason"*. In the end—forgive us for the play on words given the subject—it seems to us that Bartholomew does not intend to stir the waters.

Thomas of Cantimpré (1201–1272)

The biography of Thomas of Cantimpré (Thomas Cantimpratensis) is very similar to that of the encyclopedists, his contemporaries. He was born at Saint Pieters-Leuw near Brussels (in the Duchy of Brabant) in 1201. He began his education at Liège, where he remained eleven years, then moved to the Abbey of Cantinpré, where he became canon regular of St. Augustine. Fifteen years later, he entered the order of St. Dominic at Leuven, afterwards studied at Cologne under the tutelage of Albert the Great, then went to Paris to perfect his studies. He went back to Leuven in 1240 and there was made professor of philosophy and theology and wrote the work of which we shall speak. He died at Leuven in 1272.

Thomas is the author of many writings: besides some hagiographies, his most famous works are the encyclopedia *De Natura rerum* (1241–1255) and *Bonum universale de apibus* (1257–1263), where he represents an ideal of Christian life under the image of the life of bees. The work *De natura rerum*, whose writing required fifteen years, was the result of reference and reworking of very many works, both known for some time and of recent translation, and where

---

infrigidatum infrigidat orbem Saturni sibi magis proximus et vicinum. Sed qualiter istud posset fieri rationabiliter, non est perspique ratione retentibus manifestum. Nam cum aquosa substantia ratione utriusque qualitatis suae scilicet humiditatis et frigititatis, coelesti sbstantiae penitus sit contraria, non est bene liquidum philosophantibus, qualiter inter corpora tam disparia possit unitas aut concordia aliqualiter conveniri. Et tamen scriptum est Job. Qui facit concordiam in sublimibus suis. Ideo alio modo sentiunt et opinantur moderni, qui interiora philosophiae spectamina profundius, ut arbitrer, sunt scrutati. Dicit enim Alexander, quod aquae illae, quae super coelo sunt, non sunt ibi positae ut frigidae et fluxibiles et humidae, vel etiam sicut solidae, congelatae, et ponderosae. Istae enim proprietates sunt etiam inter se habentes contrarietatem, et sibi mutuo repugnantes, sed potius per ordinationem divinae sapientiae aquae illae super firmamentum sunt sub nobilissima naturae suae conditione divinitus collocatae, prout naturae coelesti maximae sunt propinquae. Et haec quidem est naturae proprietas perspiquitatis et transparentiae, quae principaliter et substantialiter invenitur in natura aquae, ratione cuius habet convenientiam et cum coelo empyreo et etiam cum firmamento. Et ideo posuit Dominus aquas inferius sub ratione frigidi et humidi, cum aliis conditionibus necessariis ad generationem et corruptionem, sed easdem posuit in ratione perspiqui prout fuit necessarium ad universitatis conservationem. et ideo dicit coelum esse dictum aqueum et crystallinum ratione mobilitatis et perspiquitatis. Est enim perspicuum ad modum crystalli, a superiori coelo scilicet ab empyreo, lucem vel fumositatis plenitudinem recipiens, receptum ad inferiora diffundens. Et ideo dicitur coelum quasi nobis invisibile et occultum, crystallinum, non quia durum sicut crystallus. Sed quia uniformiter est luminosum et perspicuum. Aqueum autem dicitur, quemadmodum aqua ex sua subtilitate et mobilitate movetur, et illud motum movet coelum proximum, et illud ulterius movet quod sibi est magis propinquum. Et ideo illud coelum quod movet inferiora, inferiorum mobilium praecipue est conservativum, ut dicit Alexander (our Eng. trans.).

[73] For this, see Iolanda Ventura, "Bartolomeo Anglico e la cultura filosofica e scientifica dei frati nel XIII secolo: aristotelismo e medicina nel *De proprietatibus rerum*", in *I Francescani e le Scienze*, Atti del XXXIX Convegno internazionale, Assisi, 2011 (Spoleto : Fondazione Centro italiano di studi sull'alto Medioevo, 2012), pp. 49–140. The authoress studies in particular the part of the work regarding zoology and human anatomy.

Aristotle's authority is greatly recognized.[74] The work underwent two writings: to the first, consisting of nineteen books, the author then added a twentieth book (*De ornatu celi et eclipsibus solis et lune*). It is the first chapter of this book (which consists of twenty-three chapters) which we are interested in and of which we report the initial part: "After the end of our work, we also add to it a twentieth contribution, not taken from a compilation by ourselves but as a necessary supplement to the preceding work itself. Nevertheless we added some things and removed some others and did not correct anything. It is our intention to write about the ornamentation of the heaven and the motion of the stars and of the planets; and it is fundamental in order to grasp the distribution of the signs on the sphere, to know the reason of the eclipse of the sun and of the moon. Here, first of all, one must speak of the ornamentation of the heaven. Truth to tell, the ornamentation of the heaven is anything which one sees over the moon, clearly both the fixed stars and the wandering ones. Some might ask: is there anything over the moon, besides the ether and the stars and the spirits we have already dealt with, or are there congealed waters, above which there are other waters? However, some say that above the ether there are congealed waters, which appear to our eyes as a taut skin, above which there are waters; this is confirmed by the authority of the divine page which says: *He put the firmament amidst the waters;* and again: *Divided the waters which are under the firmament from the waters which are above the firmament.* If in that place there are waters, then there is something weighty and solid. But the right place of the weighty and solid things is the earth. If in addition there are congealed waters, they are either combined with the fire or not. If they are combined with the fire, since this is warm and dry whereas the congealed waters are cold and humid, then two opposing things are combined without having anything between them. Therefore there is never harmony, but rather repulsion from opposing things. Further, if the congealed waters were combined with the fire, either they would evaporate because of the fire or they would extinguish the fire. Therefore, if the fire and the firmament remain in their place, there are no congealed waters combined with the fire. If they are not combined, something exists between them and the fire. But what? Is it an element? Hence visible? Why then one does not see it? Therefore it is indubitable that there are no congealed waters. But in truth, and in accordance with the Catholic doctrine, we believe in what the Church maintains, that is, that there are waters above the firmament, even if our reason cannot go so far as to grasp it".[75]

---

[74] For this, see Mattia Cipriani, "*Questio satis iocunda est*: Analisi delle fonti di *questiones* et *responsiones* del *Liber de natura rerum* di Tommaso di Cantimpré". *Rursus* 11 (2017). http/journals.openedition.org/rursus/1330.

[75] Thomas of Cantimpré, *Liber de Natura Rerum. Editio princeps secundum Codices manuscriptos* (Berlin-New York: Walter De Gruyter-Berlin-New York, 1973), Incipit liber XX, p. 415: *De Ornatu Celi et eclipsibus Solis et Lune. I: Post finem laboris nostri vicesimam quoque editionem apponimus, sed hanc non tanquam ex nostra compilatione, sed tanquam necessariam ipsi operi precedenti. Addidimus tamen aliqua et quedam subtraximus atque nonnulla correximus. Est autem scribentis intentio de ornatu celi et motu syderum atque planetarum; et valet multum ad intelligendam speram in distributione signorum, ad agnoscendam rationem eclipsis solis sive lune. Hinc primum de ornatu celi dicendum est. Ornatus vero eius est quicquid supra lunam videtur, scilicet stelle tam infixe quam erratice. Sed queret aliquis. estne super lunam nisi ether et stelle et pretaxati spiritus, an ibi sunt aquae congelatae, super quas sunt alie aquae? Dicunt tamen quidam super ethera esse aquas congelatas, que in modum pellis extense oculis nostris occurrunt, super quas sunt aque; confirmantes hoc auctoritate divine pagine que ait: Posuit firmamentum in medio aquarum; et iterum: Divisit aquas que sub firmamento sunt ab hiis, que sunt super firmamentum. Si ibi sunt aque, ergo aliquid ponderosum et grave. Sed proprius locus ponderosorum et gravium est terra. Si iterum aque congelate ibi sunt, coniuncte sunt igni vel non. Si igitur coniuncte sunt cum igne, qui cum calidus sit et siccus, aqua congelata frigida et humida, contrarium sine medio suo contrario coniunctum est. Numquam ergo ibi est concordia, sed contrariorum repugnantia. Amplius: si aqua congelata coniuncta est igni, vel dissolveretur ab igne vel extinguet ignem. Cum ergo ignis et firmamentum remaneant, non sunt aque congelate igni coniuncte. Si coniuncte non sunt, aliquid inter eas et ignem est. Sed quid? Est elementum? Sed nullum superius igne factum est ex elementis. Visibile ergo? Unde ergo non videtur? Restat ergo ibi non esse*

Our comment is that Thomas's presentation of the problem is flawless. The argument against the existence of the waters above the firmament is explained with clarity, by summarizing the conclusions of his predecessors, and his conclusion is very clear: the Bible asserts this and therefore this is undoubtedly true, but we are not able to understand it. Therefore the reader is exactly informed about how the things stand and the encyclopedist has done his duty of supplying a scrupulous account of them.

Arnold of Saxony (?–?)

About Arnold of Saxony (Arnoldus Saxo) truly very few facts are known; even the addition of *Saxony* to the name *Arnold* is somehow uncertain. But it must be said that the discovery of the existence of his encyclopedia is relatively recent (1855!), thanks to the German scholar Valentin Rose who was dealing with the studies of Aristotle on stones.[76] Obviously, following the discovery several researches have been carried out and, at least from the point of view of the works that can certainly be attributed to him, the situation is now more definite.[77]

As Isabelle Draelants says, "Nearly everything about Arnold of Saxony can only be known through his work, whose handwritten tradition and ancient testimonies refer back to a Germanic space".[78] It has also been ascertained that he worked in the thirteenth century, and it is believed that his encyclopedia, about which we now shall speak, can be dated between 1225 and 1240. Entitled *De Floribus rerum naturalium*, the work is divided in five parts arranged as independent books, each one with an own prologue. Their titles are: *De coelo et mundo, De naturis animalium, De virtutibus lapidum, De virtute universali, De moralibus*. The first part, *De coelo et mundo*, is in its turn divided in five books, the second of which has thirteen chapters (*De natura stellarum, De motibus astrorum, De natura planetarum, De motibus et iudiciis planetarum, De eclipsi solis et lune*, etc.).

We have reported all these titles in order to explain our expectation of finding (in the part *De coelo et mundo*) the question of the waters above the firmament. Instead, despite the titles of the chapters being like those in the encyclopedias of his contemporaries, the question does not appear. The explanation is simple. As Isabelle Draelants shows, the sources of inspiration of the author must not be looked for in Genesis, but in Plato (*Timaeus* and its commentaries) and in Aristotle.[79] Hence Arnold's encyclopedia is a "lay" encyclopedia, with what this adjective may be worth. It is certainly an exception in the thirteenth century and that's why we have dwelt on it.

Vincent of Beauvais (1184–1264)

We agree with Lynn Thorndike, when he says: "Of medieval encyclopedists and compilers Vincent of Beauvais may be ranked as chief by reason of his *Speculum Maius*, which really consists of three voluminous 'Mirrors', the *Speculum Naturale*, with which we shall chiefly

---

aquas congelatas. Sed vere et catholice profitemur secundum quod ecclesia tenet, aquas esse super firmamentum, et si ratio nostra haec non valeat indagare (our Eng. trans.).

[76] V. Rose, "Aristoteles de lapidibus und Arnoldus Saxo", *Zeitschrift für deutsche Altertum*, t. 18 (1855), pp. 321-455.

[77] In this regard, see the fundamental Isabelle Draelants, Un encyclopédiste méconnu du XIIIe siècle: Arnold de Saxe (Ph.D. thesis, Université catholique de Louvain, 2001). https://tel.archives-ouvertes.fr/tel-00700745, accessed 13.01.2020.

[78] I. Draelants, *Un encyclopediste méconnu*, op. cit., p. 6: *A peu des chose près, Arnold de Saxe ne peut être connu q' à travers son oeuvre, dont la tradition manuscrite et les témoinages anciens renvoient à un espace germanique* (our Eng. trans.).

[79] See I. Draelants, *Un encyclopediste méconnu*, op. cit. pp. 130-132.

concerned, and the *Speculum Doctrinale* and *Speculum Historiale*".[80] We too shall deal with the *Speculum Naturale*, but before, as usual, we shall give some biographical information about the author.

Vincent of Beauvais (Vincentius Bellovacensis) was born around 1184, perhaps at Beauvais, since he speaks of himself as "Vincent of Beauvais of the Order of Preachers". He became a Dominican Friar at Paris (before 1220), where he was also a preacher in the court of King Louis IX, who appointed him "lector" of the school of the Cistercian monastery of Royamont-sur-Oise. He was also the tutor of the king's son and, in relation to that function, he wrote the treatise *De eruditione filiorum regalium* (1250–1252), at the request of Queen Margaret. He died at the abbey of Royamont in 1264.

As mentioned above, his most important work was the *Speculum maius*, whose writing took up several years around the half of the century. The work consists of three parts: *Speculum naturale* (32 books) with notions of natural sciences; *Speculum doctrinale* (17 books), a collection of notions on arts and doctrine for elevating man through knowledge; and *Speculum historiale* (31 books), a summary of the history of humankind from Adam to 1250. To these three *Specula*, a fourth, *Speculum morale*, was later added, but it is of uncertain attribution. As a whole, the work, though being in general devoid of original elaboration, turned out notably useful since, by reporting excerpts of many works (even Arab), it contributed to the diffusion of culture in the Latin world.

Let us speak specifically of the *Speculum naturale*.[81] The subject of the work is outlined by Vincent himself in the Prologue (Chap. XVII): "So the first part is the foundation of the sacred history from the very beginning of creation to the Sabbath rest. And then, the path will be those that concern the nature of Heaven and the world".[82] That is, it was to be a sort of Hexaemeron supplemented by notes and explanations about the nature of the heaven and the world. Actually, its consideration of nature follows the order of the six days of the creation, but the host of "scientific" data is such to obscure the underlying Biblical plan. In fact the *Speculum naturale* consists of 32 books, each one with many chapters. The book we are interested in is the third, which begins with the works of the second day (*De opere secundae diei)* and consists of 105 chapters. Several chapters are devoted to the firmament, to time and its nature, to the Trinity and, finally, to the waters above the firmament.

The method used in dealing with the various subjects is shaped according to that of *quaestiones et responsiones*. In fact, he first shows the opinions of the various authors (Fathers of the Church and philosophers), then the refutations (*objecta in contrarium*). Often he also inserts his own opinion (under the wording *Auctor*) but as a rule he does not add anything of importance.

Let us see the case of the waters above the firmament. He begins to speak of them in Chap. XC (*De caelo aqueo sive crystallino*) quoting the opinion of Augustine and continuing with those of William of Conches, Ambrose, John Damascene, Albert the Great, Basil, but also Aristotle, Ptolemy, Averroes, and others. Following William of Conches, he rejects the opinion of Bede by choosing the *Probabilior modernorum sententia de hac materia* (Chap. XCIV). The adjective *probabilior* recalls Abelard, but it seems to us that Vincent, rather than making the question clear, tends to confuse it and leave the reader to extricate himself from the confusion

---

[80] L. Thorndike, *History of Magic and Experimental Science*, op. cit. p. 457.

[81] The Latin text to which we shall refer is: Vincent of Beauvais, *Speculum Quadruplex: Naturale, Doctrinale, Morale, Historiale* (Douai: Baltazaris Belleri, 1624; rpt. Graz: Academische Druk-u Verlagsunstalt, 1964-65).

[82] Vincent of Beauvais, *Speculum : Igitur primae partis fundamentum est historia sacra ab ipso principio creationis rerum usque ad requiem sabbahti. Cui etiam interferantur ea quae pertinent ad naturam coeli et mundi.*

of the various opinions; this gives us cause to long for the conciseness and clearness of Thomas of Cantimpré. The theological framework of the work often seems to override the "scientific" explanations. Perhaps Vincent was convinced that the readership of his encyclopedia did not consist of people who looked for information about the natural sciences, but believers who wanted to be reassured that the natural events they assisted belonged to the divine plan.

2.2.7 Robert Grosseteste (1168/1175–1253)

After the encyclopedists, we now deal with the philosophers who in the thirteen century wrote commentaries on Genesis and dealt in particular with the question of the waters above the firmament. One of the most important among them is undoubtedly Robert Grosseteste who, even though having carried on a feverish activity in the Church of his time and finished his life as bishop of Lincoln, also devoted himself, in addition to theological questions, to scientific questions and to the definition of the scientific method. Precisely for this reason he has been considered the forerunner of the modern science[83] and of the Galilean method.[84]

The biographical notes on the first part of the life of Grosseteste are uncertain and fragmentary, beginning with the date of his birth is not certain, which ranges from 1168 to 1175. He was born in the country of Suffolk and it seems that he studied the arts at Oxford. Afterwards, he was in various places (Hereford, Leicester, Oxford, and perhaps also the France) and held important positions. He also was very active in defending the English clergy, both against the king and against the pope. In 1235 he was made bishop of Lincoln, where he remained until his death.

The work of Grosseteste with which we shall deal is the *Hexaemeron*.[85] The work consists of eleven parts (*Particulae*), preceded by a long proem, in which the author relates various historical facts and philological considerations in the manner of Isidore. As Dales and Gieben, editors of the critical edition, say: "Hexameral literature has played an important role in Christian thought since the patristic age, and the re-introduction of ancient thought into western Europe during the twelfth and the thirteenth centuries inspired new writings on the old themes, designed to meet the needs and answer the questions of a new age. A work of critical importance in this tradition was the *Hexaëmeron* of Robert Grosseteste".[86]

It is thought that Grosseteste wrote this work in the period which ranges from 1228 to 1235; in the same period he also devoted himself to studying Greek and began the translation of Aristotle's works. At the beginning of the work (*Particula prima*), we find a series of sentences which can be defined "of principle". First, "Each science, each kind of wisdom has a matter and a subject on which its attention is turned".[87] Then, after having clarified that the wisdom of which he speaks is theology, he adds: "…the subject of this wisdom is neither known in its own right, nor received through science. it is only accepted and believed through faith. Nor can it be

---

[83] See, for instance, A. C. Crombie, *Robert Grosseteste and the origin of Experimental Science, 1100–1700* (Oxford, Clarendon Press, 1953).

[84] We have touched on this subject in our book: Dino Boccaletti, *The Shape and Size of the Earth*, op. cit., pp. 105-107.

[85] See the critical edition: Robert Grosseteste, *Hexaemeron*, eds. Richard C. Dales and Servus Gieben (Oxford University Press, 1982) and the English translation: Robert Grosseteste, *On the Six Days of Creation*, trans. C. F. J. Martin (Oxford University Press,1996).

[86] Grosseteste, *Hexaemeron*, ed. Dales-Gieben, op. cit., p. 11.

[87] Grosseteste, *Hexaemeron*, op. cit., I, I, 1: *Omnis scientia et sapientia materiam habet et subiectum aliquod, circa quod eiusdem versatur intentio. Unde et hec sapientia sacratissima, que theologia nominatur, subiectum habet circa quod versatur* (Eng. trans. C. F. J. Martin, from Robert Grosseteste, *On the Six Days of Creation: A Translation of the Hexaëmeron*, op. cit., p. 47).

understood, unless it is first believed".[88] And again: "Things believable are of two kinds. Some things are believable because of the likelihood of the things themselves; others because of the authority of the one who speaks. In this wisdom believability on account of the likelihood of things is an accident. For the things that in this wisdom are believable, properly speaking, are believable on account of the authority of the one who speaks. Hence, since in this text there is no difference in the authority of one who speaks—since it is all the authority of God who speaks 'by the mouth of his holy prophets, who have been from the beginning'—there is no difference in the believability of the things to be believed in this writing or scripture...".[89] And, finally: "The species of this world, in so far as regards the way they are now governed, have the certainty of sense and science. But in so far as regards the ordering in which they were created, they cannot be grasped at first except by faith. So the creation of the sensible world, on account of the way in which the world is imaginable and graspable by the external senses of the body, should be told in the opening part of the Scripture. This is in order that anyone, even among the inducted, may be able to grasp a story of this kind easily, through his imagination and through the images of corporeal things, and grow stronger in faith through the authority of the one who speaks".[90] With regard to his future readers, he will say below, when speaking about the heaven (or the heavens): "...in our modest way, we take care to put them in here, so that the humble reader—for we do not write these things for the wise and the perfect—may have to hand some material, so that he can easily adapt the heaven to the meanings given above, and to be mentioned elsewhere, according to these properties and likenesses".[91]

His insisting on the nature of the audience of the book is also noticed by the editors of the critical edition: "His references to the "reader" rather than to the "auditor" and his statement that he is not *writing* for sages and perfecti make it clear that this work was composed to be read as a book."[92] That is, the work was purposely written as book to be read as a guide to grasp Genesis, and not as a guide for sermons.

Grosseteste even takes care to account for the repetitions in the text, done according to him in order that the children can also understand: "But Scripture, which is kind to little children, reminds us more than once of God's single utterance, because the minds of little children do

---

[88] Grosseteste, *Hexaemeron*, op. cit., I, II, 1: *Sive igitur huius sapiencie subiectum sit istud unum quod diximus sive Christus integer, istius sapiencie subiectum neque per se notum est, neque per scientiam acceptum, sed sola fide assumptum et creditum. Nec posset esse intellectum, nisi prius esset creditum* (Eng. trans. C. F. J. Martin, op. cit, p. 48).

[89] Grosseteste, *Hexaemeron*, op. cit., I, II, 2: *Credibilia autem duplicia sunt: quedam enim sunt credibilia propter ipsarum rerum verisimilitudinem, quedam vero propter dicentis auctoritatem. In hac autem sapiencia credibilitas ex rerum verisimilitudine accidens est. Nam que in hac sapiencia proprie credibilia sunt, a dicentis auctoritate sunt credibilia. Unde, cum in hac scriptura indifferens sit dicentis auctoritas, Dei videlicet loquentis* per os sanctorum, qui a seculo sunt prophetarum eius, *indifferens est et credendorum in hac scriptura credibilitas* (Eng. trans. C. F. J. Martin, op. cit, p. 49).

[90] Grosseteste, *Hexaemeron*, op. cit., I, II, 3: *Species autem huius mundi, secundum quod nunc gubernantur, habent sensus et sciencie certitudinem. Secundum ordinem vero quo creabantur, non accipiuntur primo nisi per fidem. Mundi igitur sensibilis creacio, per modum quo mundus ymaginabilis est et per corporis exteriores sensus apprehensibilis, in primordio huius scripture debuit enarrari, ut quivis eciam rudis huiusmodi narracionem facillime possit per ymaginacionem et rerum corporalium ymagines apprehendere, et per dicentis auctoritatem in fide firmare* (Eng. trans. C. F. J. Martin, op. cit., p. 49).

[91] Grosseteste, *Hexaemeron*, op. cit., III, XVI, 1: *...pro modulo nostro hic interserere curemus, ut habeat lector parvulus in promptu - non enim sapientibus et perfectis ista scribimus - ex quibus proprietatibus et similitudinibus possit faciliter aptare celum supradictis significatis et alibi dicendis* (Eng. trans. C. F. J. Martin, op. cit, p. 118).

[92] Grosseteste, *Hexaemeron*, op. cit., p. XII.

not understand at one grasping of it that God by his single word spoke and made everything".[93] Hence, for each creative act, the announcement must be repeated in order to facilitate comprehension of it.

The question of the waters above the firmament is a subject which recurs in several parts of the work (I, III, V, IX, X), together with the discussion on the nature of the firmament. In essence, Grosseteste thinks that everything has already been told about the firmament and the waters and therefore what the reader of Genesis must do is: 1) believe the letter of the text; and 2) consider the interpretations suggested by the Fathers of the Church, which are all valid. He also tries to formulate his interpretation of Genesis in a methodical manner, even by establishing rules to be followed rigorously. Let us see them: "1. So the first literal sense of the beginning of Scripture has to do with the successive creation in time of bodily and visible things, of heaven and earth and their visible adornment. It would not be right for the first literal meaning to speak expressly of the creation of the angels, since they are substances which we can grasp only with our intellect, and do not belong with those easily-grasped substances with which Scripture opens. Nor would it even be right for the first literal sense to refer to the creation from nothing of unformed, non-sensible first matter. But it is right for the first literal, imaginable sense to point to the world which is grasped only by the intellect, in order that in this opening there should be meat for the perfect to eat and milk for babes to suck. 2. Hence the first literal sense of 'created world' refers to 1) the uncreated archetypal world, that is, to the eternal and unchangeable ideas of the created world that are in the mind of God. The literal, imaginable sense also refers to 2) the establishment of the angels and to the knowledge, in the mind of the angels, of the world that had to be created. It refers, too, to 3) the creation of the primordial matter and form out of nothing, and to 4) the ordered establishment of the sensible world out of them. It refers, too, in an allegorical way, to 5) the ordering of the Church, and in a tropological way to 6) the perfecting of the soul by faith and morals. There are, then, to sum up, six different ways of understanding and expounding this opening which deals with the creation of the world in six days. Perhaps these six ways of expounding are hinted at by the six days and their works".[94]

Regarding an allegorical interpretation of the works of the second day: "4. The second day could be the created intelligence of the angels. The possibility of turning that belongs to its free choice, which is to remain fixed for ever, is like the turning firmament. The water above the

---

[93] Grosseteste, *Hexaemeron*, op. cit., III, I, 1: *...Scriptura, condescendens parvulis, meminuit pluries unicam Dei dictionem, quia mentes parvulorum non unica apprehensione comprehendent, quod Deus unico verbo loquitur et facit omnia* (Eng. trans. C. F. J. Martin, op. cit, p. 102).

[94] Grosseteste, *Hexaemeron*, op. cit., I, III, 1-2: 1. *Primus igitur sensu litere principii huius scripture est de creacione temporali et successiva corporalium et visibilium celi et terre, et visibilis ornatus eorum. Nec debet, prima litere significacio expresse sonare creacionem angelorum cum sint substancie solum intelligibiles, nec a communitate eorum quibus hec scriptura proponitur in principio comprehensibiles; nec debet eciam primus litere sensus signare materie prime insensibilis et informis ex nichilo creacionem. Debet tamen in primo sensu litere ymaginabili de creacione mundi, mundus intelligibilis designari, ut habeat in hoc principio perfectus quod comedat, et parvulus quod sugat.*
2. Grosseteste, *Hexaemeron*, op. cit., *Unde per sensu primum literalem mundi creati, signatur mundus increatus archetipus, id est eterne et incommutabiles raciones in mente divina mundi creati. Signatur eciam per literalem sensum ymaginabilem angelorum condicio et mundi creandi in mente angelica cognicio. Signatur quoque materie et forme primordialium ex nichilo creacio, et mundi sensibilis ex illis primordialibus ordinata condicio. Signatur quoque allegorice ecclesie ordinacio, et tropologice per fidem et mores anime informacio. Sunt igitur in summa huius primordii de creacione mundi in sex diebus sex intellectus et exposicionum modi; qui forte senarius exposicionum per sex dies et eorum opera designantur* (Eng. trans. by C.F.J. Martin, op. cit., p. 50).

firmament is the change of their intelligence for the better through success, and the water below the firmament is the change of their intelligence for the worse through failure".[95]

Let us get now to the heart of the matter we are interested in: "Then, by God's word, the firmament was made from these waters, the firmament where we now see the lights of heaven. The place and position of this firmament is far below the heaven that was made at the beginning. So part of the primordial waters was left above the firmament, and filled the whole space above it to the first heaven, while part was left below the firmament, and filled the whole space from it down to earth beneath. All this was brought to completion on the second day, a day not yet brought about by the sun, but by the light that was first made, dividing up day and night by its giving light and its bringing its light underneath. On the third day, too, which was brought back by that light, the waters that were under the firmament were gathered together: that is, they were condensed out of the vaporous waters and became the dense waters which we now see in the sea and the rivers. And they were collected into the hollows of the earth which are now the places of seas and rivers. But when the vaporous waters were condensed into denser waters, then since that which has been made dense occupies less space, and since no space can stay empty, part of the vaporous waters therefore stayed where it was, rarefied, and occupying a greater space, to match the lesser space occupied by the part that had been condensed. Thus, at the same time as the waters were gathered together air came into existence, since to rarefy vaporous waters is to produce air. If the rarefaction and thinning down is very great, then even fire and flame are produced. Now, when the waters were gathered together into seas and rivers, just as we now see them, and the air filled this place above the earth, by the word of God the earth brought forth the living things that grow in it as we now see them, each one of which was to propagate its own like according to its kind by its seed and its fruit. On the fourth day were made the lights and the stars which we now see in their place in the firmament, by whose movement and giving of light day and night came to be separated, and periods and measurement of time are shown to us. On the fifth day the word of God brought forth from the gathered waters the creeping things of the waters and the flying creatures of heaven, and God's command joined them in pairs to begin to breed their future offspring. On the sixth day the earthy animals were brought forth from the earth, and man was made in the image and likeness of his maker. That the world and its adornment were made in this order, according to this order of the days of time, is maintained by Josephus, Bede, Basil, Ambrose and Jerome, and other authorities, even though they seem to disagree on several other matters that have to do with the literal exposition of the text. This we will show, with God's help, later on".[96]

---

[95] Grosseteste, *Hexaemeron*, op. cit., I, III, 4: *Secundus autem dies potest esse creata intelligentia angelica, cuius vertibilitas arbitrii liberi in perpetuum mansuri est sicut volubile firmamentum, super quod aqua superior est mutabilis eius per profectum in melius, et aqua inferior est mutabilis eius per defectum in peius* (Eng. trans. by C.F.J. Martin, op. cit., p. 50-51).

[96] Grosseteste, *Hexaemeron*, op. cit., I, VIII, 2: *Deinde Verbo Dei de dictis aquis factum est firmamentum in quo nunc videmus celi luminaria; cuius firmamenti situs est locus multo est inferior celo in principio facto. Remanserunt igitur de dictis primordialibus aquis pars supra firmamentum, que totum undique replevit spacium a firmamento superius usque ad primum celum, et pars sub firmamento que totum undique replevit spacium a firmamento usque ad terram deorsum. Istudque completum est in die secundo, quem nondum sol fecit sed lux primo facta, diem et noctem dividens sui illustracione et luminis subductione. Tercio quoque die, quem reduxit lux dicta, aque que erant sub firmamento congregate sunt, hoc est, condensate ex vaporalibus in aquas spissas, quales nunc in mari et fluminibus videmus; collecteque sunt in terre cavitatibus, que nunc sunt loca marium et fluminum. Ipsa vero condensacione vaporalium aquarum in aquas spissiores, cum inspissatum occupet minorem locum nec possit remanere locus vacuus, pars aquarum vaporalium in tantum remansit subtiliata et occupans maiorem locum quanto reliqua pars fuerat condensata et occupans minorem locum. Sic quoque simul cum aquis congregatis fiebat aer, quia vaporalis aque rarefaccio est aeris generacio; et si fit subtilacio et rarefaccio multa, generabitur eciam ignis et flamma. Hac autem colleccione aquarum facta in maribus et fluminibus quemadmodum*

We have reported this long excerpt in its entirety because it is substantially a summary of Grosseteste's interpretation of Genesis (he even quotes the authors he considers to be authorities). What we can draw from this, for the moment, is that: 1) the place and position of the firmament is far below the heaven that was made at the beginning; 2) the primordial waters had a vaporous nature and because of this from the part of them which will remain under the firmament are obtained (through condensation) the sea and the rivers and, by subsequent rarefaction, air and fire (the earth was existing from the beginning). Thus was obtained the integration of the spheres of Empedocles' four elements into the biblical process of creation; 3) for the moment, no mention is made of the first part of the eighth verse, "*Vocavitque Deus firmamentum caelum*".

At least for the moment, Grosseteste uses the substantive *caelum* with different meanings. As we have remarked above, the certainties of Grosseteste are founded on the authorities. In fact, afterwards, he confirms: "But someone might doubt whether the heaven—which according to the first literal sense mentioned above, was said to enclose all the other bodies in the world—is above the firmament which we are told of as having been made on the second day, and whether the waters in the midst are above the firmament and below this heaven. That this heaven, taken in the literal sense, is above the firmament made on the second day, and above the waters that are above the firmament, we can gather from the authority of Jerome, of Strabus, of Bede, of John Damascene, and of Basil".[97]

However, it may happen that the authorities don't always agree. For example: "But Josephus and Gregory of Nyssa, and some other commentators, differ from the authorities cited above, and think that this heaven is nothing different from the firmament created in the second day".[98] In that case, Grosseteste takes a stand: "It is not for me to decide either way on this dispute between authorities. But if this first heaven is something different from the firmament created on the second day, it seems to be something unmoveable".[99] He goes on to explain the reason for his opinion.

Still, on the nature of the firmament not all has been said, and Grosseteste also applies to the authority he considers the most "authoritative", Augustine, who, as we know, dealt with this

---

nunc videmus, et aere replente regionem hunc superiorem, Verbo Dei produxit terra qualia nunc videmus terrenascencia, quorum unumqodque per semen et plantacionem sibi simile secundum speciem produceret. Quarto autem die facta sunt luminaria et stelle que nunc in frmamento collocata conspicimus, quorum motu et illuminacione fiunt divisim dies et nox, et signantur nobis temporum spacia et dimensiones. Quinto vero die ex aquis congregatis producta sunt Verbo Dei aquarum reptilia et celi volatilia, copulataque sunt Dei iusione ad permixtionem iam sobolis profuture. Sexto autem die producta sunt de terra terrena animalia, et homo conditus ad ymaginem et similitudinem sui conditoris. Quod autem hoc modo successive per ordinem dierum temporalium formatus sit mundus et eius ornatus, astruunt Iosephus, Beda, Basilius, et Ambrosius, et Ieronimus, et nonulli auctores alii, licet in plerisque aliis que pertinent ad exposicionem literalem dissentire videantur, ut inferius, Deo adiuvante, plenius manifestabimus (Eng. trans. by C. F. J. Martin, pp. 55-56).

[97] Grosseteste, *Hexaemeron*, op. cit., I, XVI, 1: *Potest autem hic dubitari an celum, quod secundum primum literalem sensum supradictum est omnia cetera mundi corpora circumdare, sit supra firmamentum quod secundo die factum commemoratur, et aque sint intermedie superiores firmamento, et inferiores hoc celo. Quod autem hoc celum secundum literalem intellectum sit superius secundo die facto firmamento et aquis super firmamentum, haberi potest ex auctoritatibus Ieronimi, Strabi, et Bede, et Iohannis Damasceni, et Basili* (Eng. trans. by C. F. J. Martin, p. 71).

[98] Grosseteste, *Hexaemeron*, op. cit., I, XVI, 3: *Iosephus autem et Gregorius Nisenus et quidam alii expositores, econtrario predictis auctoribus, putant hoc celum et firmamentum secundo die factum non esse diversa* (Eng. trans. by C. F. J. Martin, p. 74).

[99] Grosseteste, *Hexaemeron*, op. cit., I, XVII, 1: *Horum autem auctorum controversiam non est meum determinare, sed si celum istud primum sit aliud a firmamento secundo die creato, videtur quod illud sit immobile* (Eng. trans. by C. F. J. Martin, p. 74).

subject in the *De Genesi ad litteram*. In fact, Grosseteste says: "However, in his [i.e., Augustine's] opinion, and in that of other commentators, the word 'firmament' is more correctly understood to mean the heaven in which the stars are placed, above which waters are truthfully said to be. Augustine tries to show that this is not improbable: for if the waters which we see can be divided into so many very small particles and be made so subtle and—by the power of heat being impressed into them, or by some other means—can made so light as to be suspended above our air in the clouds, in a vaporous state, by parity of reason the same waters, divided up into into yet smaller particles, with greater subtlety and lightness, in proportion as the place above the firmament is higher than the place of the clouds, can be suspended up there. And if the power which makes them subtle and lightens them is a permanent and enduring power, then they can stay up there constantly".[100] He follows this with the already well known argument of the coldness of Saturn, but he emphasizes that the differences of opinion does not mean that any of them are mistaken: "But what is the point of repeating all these different opinions, since they cannot all be true together? Well, for this reason: that we may know the possible ways in which it may come about that there really are waters located above the firmament, and may be able to give an answer to those who try to prove that there cannot be any waters above the heavens, and show to them that there are several ways in which what Scripture says may be true. The holy commentators wrote these things not so much in order to assert one of them as true, as to show what possible ways there are for what Scripture says to be true".[101] He quotes Augustine's well known statement: "The authority of this text of scripture is greater than that of all the power of human ingenuity".[102]

However, he still feels bound to look for further certainties on the existence of the firmament: "Though we are sorry for our ignorance, we hold that the firmament exists, basing ourselves on the voice of God. The cause of the essence of the firmament, as Basil said 'is clearly declared by Scripture: "let it divide the waters from the waters"'. And there is no doubt that the firmament, both in its own right and through the stars that are placed in it—whether it moves them by some means or not—is of great help in the generation and growth of things below. For those heavenly bodies have a light which assists vital heat.[103]

---

[100] Grosseteste, *Hexaemeron*, op. cit., III, III, 1: *Verumtamen, iudicio eius et aliorum expositorum verius intelligitur nomine firmamenti celum in quo locata sunt sidera; super quod veraciter sunt aque posite. Nec hoc esse improbabile nititur Augustinus probare: quia, si aque iste, quas videmus, in tantas minucias possunt dividi et tantum subtiliari et aliqua vi impressi caloris vel alio modo in tantum levigari, ut super hunc aera possint vaporabiliter in nubibus suspendi, eadem racione eedem aque, minucius divise magisque subtiliate et levigate secundum proporcionem qua locus superior firmamento altior est loco nubium, ibidem suspendi poterunt. Et si virtus ea subtilians et levigans sit virtus fixa et manens, ibidem perseveranter manere poterunt* (Eng. trans. by C. F. J. Martin, p. 103).

[101] Grosseteste, *Hexaemeron*, op. cit., III, III, 6: *Sed quid prodest istas diversas sentencias recitasse, cum omnes simul stare non possint? Ad hoc utique, ut sciamus modos possibiles quibus potest fieri ut aque veraciter supra firmamentum existant,et possimus respondere ad illos qui nituntur probare non posse esse aquas supra celos, et ostendere eis plures modos quibus potest esse hoc quod dicit Scriptura. Sacri enim expositores scripserunt ista non tam ut horum aliquod unum esse assererent, quam ut modos possibiles ostenderent quibus hoc potest esse quod Scriptura dicit esse* (Eng. trans. by C. F. J. Martin, p. 105).

[102] Grosseteste, *Hexaemeron*, op. cit., III, III, 7: *Maior est enim huius scripture auctoritas, quam omnis humani ingenii capacitas* (Eng. trans. by C. F. J. Martin, p. 105). Cfr. Augustine, *De genesi ad litteram* II, 5.

[103] Grosseteste, *Hexaemeron*, op. cit., III, IX, 1: *Nostram itaque ignoranciam dolentes ex divina voce firmamentum esse teneamus. Cuius essencie causam, sicut dicit Basilius, "apercius declaravit Scriptura, id est, ut dividat inter aquam et aquam". Nulli eciam dubium, quin firmamentum et per se et per stellas in eo locatas, sive moveat eas aliquo modo sive non, magnum prestet iuvamentum generacioni et profectui rerum inferiorum. Habent enim illa celestia corpora lumen iuvativum caloris vitalis* (Eng. trans. by C. F. J. Martin, p. 109).

And, finally, the inevitable question: "But what is the use of the waters that are above the firmament?"[104] The natural reply is a review of the opinions in this regard expressed by the "authorities". Coming to an end, he goes on with the discussion on the heaven, involving the "humble reader": "Although, as we said above, we do not know what the heaven is or how many heavens there are, or whether anything moves the heaven, or, if the heavens are moved, how many movements we have—even so, it is relevant to try to put in some details of the properties of the heavens. Some of these we are sure about, of others less sure: but the commentators have presupposed them and they are presupposed in the mystical meanings. Hence, in our modest way, we take care to put them in here, so that the humble reader—for we do not write these things for the wise and the perfect—may have to hand some material, so that he can easily adapt the heaven to the meanings given above, and to be mentioned elsewhere, according to these properties and likenesses".[105]

Up to this point, we have tried to follow the narration of the six days of creation given by Grosseteste in his commentary to Genesis, but now we want to compare this narration with the cosmogony he drafts in another of his works. As is known, he is also the author of a series of scientific and philosophical tracts, among which one of the most quoted is that devoted to light: *De luce seu de inchoatione formarum*,[106] a short essay in which light is considered to be the first corporeal form and cosmogonic principle. Let us see two excerpts of it.

The first is on the nature of light: "The first corporeal form which some call corporeity is in my opinion light. For light of its very nature diffuses itself in every direction in such a way that a point of light will produce instantaneously a sphere of light of any size whatsoever, unless some opaque object stands in the way. Now the extension of matter in three dimensions is a necessary concomitant of the corporeity, and this despite the fact that both corporeity and matter are in themselves simple substances lacking all dimensions. But a form that is in itself simple and without dimensions could not introduce dimension in every direction into matter, which is likewise simple and without dimension, except by multiplying itself and diffusing itself instantaneously in every direction and thus extending matter in its own diffusion. For the form cannot desert matter, because it is inseparable from it, and matter itself cannot be deprived of form. But I have proposed that it is light which possesses of its very nature the function of multiplying itself and diffusing itself instantaneously in all directions. Whatever performs this operation is either light or some other agent that acts in virtue of its participation in light to which this operation belongs essentially. Corporeity, therefore, is either light itself or the agent which performs the aforementioned operation and introduces dimensions into matter in virtue of its participation in light, and acts through the power of this same light. But the first form cannot introduce dimensions into matter through the power of the subsequent form. Therefore light is not a form subsequent to corporeity, but it is corporeity itself".[107]

---

[104] Grosseteste, *Hexaemeron*, op. cit., III, X, 1: *Sed que utilitas aquarum super firmamentum?* (Eng. trans. by C. F. J. Martin, p. 100).

[105] Grosseteste, *Hexaemeron*, op. cit., III, XVI, 1: *Licet vero, sicut supra diximus, jgnorentur a nobis celi natura et celorum numerus, et an celum aliquod moveatur, et si moventur celi, qui sunt modi motuum illorum; tamen non ab re est, si celi proprietates, quasdam nobis certas quasdam vero minus certas, ab expositoribus tamen suppositas et in misticas significaciones assumptas, pro modulo nostro hic interserere curemus, ut habeat lector parvulus in promptu—non enim sapientibus et perfectis ista scribimus—ex quibus proprietatibus et similitudinibus possit faciliter aptare celum supradictis significatis et alibi dicendis* (Eng. trans. by C. F. J. Martin, p. 118).

[106] See the Latin text in: *Die Philosophischen Werke des Robert Grosseteste, Bischofs von Lincoln*, ed. Ludwig Baur (Münster, 1912), and the English translation: Robert Grosseteste, *On Light*, trans. Clare C. Riedl (Milwaukee: Marquette University Press, 1942).

[107] Grosseteste, *De luce*, op. cit.: *Formam primam corporalem, quam quidam corporeitatem vocant, lucem esse arbitror. Lux enim per se in omnem partem se ipsam diffundit, ita ut a puncto lucis sphaera lucis quamvis magna*

The second excerpt is on the cosmogonic process: "To return therefore to my theme, I say that light through the infinite multiplication of itself equally in all directions extends matter on all sides equally into the form of a sphere and, as a necessary consequence of this extension, the outermost parts of matter are more extended and more rarefied than those within, which are close to the center. And since the outermost parts will be rarefied to the highest degree, the inner parts will have the possibility of further rarefaction.

"In this way light, by extending first matter into the form of a sphere, and by rarefying its outermost parts to the highest degree, actualized completely in the outermost sphere the potentiality of matter, and left this matter without any potency to further impression. And thus the first body in the outermost part of the sphere, the body which is called the firmament, is perfect, because it has nothing in its composition but first matter and first form. It is therefore the simplest of all bodies with respect to the parts that constitute its essence and with respect to its quantity which is the greatest possible in extent. It differs from the genus body only in this respect, that in it the matter is completely actualized through the first form alone. But the genus body, which is in this and in other bodies and has in its essence first matter and first form, abstracts from the complete actualization of matter through the first form and from the diminution of matter through the first form.

"When the first body, which is the firmament, has in this way been completely actualized, it diffuses its light (*lumen*) from every part of itself to the center of the universe. For since light (*lux*) is the perfection of the first body and naturally multiplies itself from the first body, it is necessarily diffused to the center of the universe. And since this light (*lux*) is a form entirely inseparable from matter in its diffusion from the first body, it extends along with itself the spirituality of the matter of the first body. Thus there proceeds from the first body light (*lumen*), which is a spiritual body, or if you prefer, a bodily spirit. This light (*lumen*) in its passing does not divide the body through which it passes, and thus it passes instantaneously from the body of the first heaven to the center of the universe. Furthermore, its passing is not to be understood in the sense of something numerically one passing instantaneously from that heaven to the center of the universe, for this is perhaps impossible, but its passing takes place through the multiplication of itself and the infinite generation of light (*lumen*). This light (*lumen*), expanded and brought together from the first body toward the center of the universe, gathered together the mass existing below the first body; and since the first body could no longer be lessened on account of its being completely actualized and unchangeable, and since, too, there could not be a space that was empty, it was necessary that in the very gathering together of this mass the outermost parts should be drawn out and expanded. Thus the inner parts of the aforesaid mass came to be more dense and the outer parts more rarefied; and so great was the power of this light (*lumen*) gathering together—and in the very act of gathering, separating—that the outermost parts of the mass contained below the first body were drawn out and rarefied to the

---

subito generetur, nisi obsistat umbrosum. Corporeitas vero est, quam de necessitate consequitur extensio materiae secundum tres dimensiones, cum tamen utraque, corporeitas scilicet et materia, sit substantia in se ipsa simplex, omni carens dimensione. Formam vero in se ipsa simplicem et dimensione carentem in materiam similiter simplicem et dimensione carentem dimensionem in omnem partem inducere fuit impossibile, nisi se ipsam multiplicando et in omnem partem subito se diffundendo et in sui diffusione materiam extendendo, cum non possit ipsa forma materiam derelinquere, quia non est separabilis, nec potest ipsa materia a forma evacuari. Atqui lucem esse proposui, cuius per se est haec operatio, scilicet se ipsam multiplicare et in omnem partem subito diffundere. Quicquid igitur hoc opus facit, aut est ipsa lux, aut est hoc opus faciens in quantum participans ipsam lucem, quae facit per se. Corporeitas ergo aut est ipsa lux, aut est dictum opus faciens et in materiam dimensiones inducens, in quantum participat ipsam lucem et agit per vrtutem ipsius lucis. At vero formam primam in materiam dimensiones inducere per virtutem formae consequentis ipsam est impossibile. Non est ergo lux forma consequens ipsam corporeitatem, sed est ipsa corporeitas (Eng. trans. by Clare C. Riedl, op. cit., p. 10).

highest degree. Thus in the outermost parts of the mass in question, the second sphere came into being, completely actualized and susceptible of no further impression. The completeness of actualization and the perfection of the second sphere consists in this that light (*lumen*) is begotten from the first sphere and that light (*lux*) which is simple in the first sphere is doubled in the second.

"Just as the light (*lumen*) begotten from the first body completed the actualization of the second sphere and left a denser mass below the second sphere, so the light (*lumen*) begotten from the second sphere completed the actualization of the third sphere, and through its gathering left below this third sphere a mass of even greater density. This process of simultaneously gathering together and separating continued in this way until the nine heavenly spheres were completely actualized and there was gathered together below the ninth and lowest sphere the dense mass which constitutes the mass of the four elements".[108]

A comparison, even if simplistic, with the excerpt from *Hexaemeron*, I, VIII, 2 (see note 96) points out the differences, so to speak, in the ground. Whereas the text of the *Hexaemeron* obviously respects the biblical account, the cosmogony presented in *On Light* turns out to be an independent physical theory, quite different even in the sequence of events. Here the first body created is the firmament and the waters do not appear (recall that Genesis 1:2 says "*et Spiritus ferebatur super aquas*") and the water is present only as one of the four elements. In conclusion, the philosophical tract proceeds within an ambit completely different from that of the biblical commentary: the things Grosseteste says when he wears the habit of the philosopher

---

[108] Grosseteste, De luce: *Rediens igitur ad sermonem meum dico, quod lux multiplicatione sui infinita in omnem partem aequaliter facta materiam undique aequaliter in formam sphericam extendit, consequiturque de necessitate huius extensionis partes extremas materiae plus extendi et magis rarefieri, quam partes intimas centro propinquas. Et cum partes extremae fuerint ad summum rarefactae, partes interiores adhuc erunt maioris rarefactionis susceptibiles. Lux ergo praedicto modo materiam primam in formam sphaericam extendens et extremas partes ad summum rarefaciens, in extima sphaera complevit possibilitatem materiae, nec reliquit eam susceptibilem ulterioris impressionis. Et sic perfectum est corpus primum in extremitate sphaerae, quod dicitur firmamentum, nihil habens in sui compositione nisi materiam primam et formam primam. Et ideo est corpus simplicissimum quoad partes constituentes essentiam et maximam quantitatem, non differens a corpore genere nisi per hoc quod in ipso materia est completa per formam primam solum. Corpus vero genus, quod est in hoc et in aliis corporibus, habens in sui essentia materiam primam et formam primam, abstrahit a complemento materiae per formam primam et a diminutione materiae per formam primam. Hoc itaque modo completo corpore primo, quod est firmamentum, ipsum expandit lumen suum ab omni parte sua in centrum totius. Cum enim sit lux perfectio primi corporis, quae naturaliter se ipsam multiplicat a corpore primo, de necessitate diffunditur lux in centrum totius. Quae cum sit forma tota non separabilis a materia in sui diffusione a corpore primo, secum extendit spiritualitatem materiae corporis primi. et sic procedit a corpore primo lumen, quod est corpus spirituale, sive mavis dicere spiritus corporalis. Quod lumen in suo transitu non dividit corpus per quod transit, ideoque subito pertransit a corpore primi caeli usque ad centrum. Nec est eius transitus, sicut si inteligeretur aliquid unum numero transiens subito a caelo in centrum –hoc enim forte est impossibile –, sed suus transitus est per sui multiplicationem et infinitam generationem luminis. Ipsum ergo lumen a corpore primo in centrum expansum et collectum molem existentem infra corpus primum congregavit; et cum iam non potuit minorari corpus primum, utpote completum et invariabile, nec potuit locus fieri vacuus, necesse fuit, ipsa in congregatione partes extimas molis extendi et disgregari. Et sic proveniebat in intimis partibus dictae molis maior densitas, et in extimis augmentabatur raritas; fuitque potentia tanta luminis congregantis et ipsa in congregatione segregantis, ut ipsas partes extimas molis contentae infra corpus primum ad summum subtiliarent et rarefacerent. Et ita fiebat in ipsis partibus extimis dictae molis sphaera secunda completa nullius impressionis ultra receptibilis. Et sic est complementum et perfectio sphaerae secundae: lumen quidem gignitur ex prima sphaera, et lux, quae in prima sphaera est simplex, in secunda est duplicata. Sicut autem lumen genitum a corpore primo complevit sphaeram secundam et intra secundam sphaeram molem densiorem reliquit, sic lumen genitum ex sphaera secunda sphaeram tertiam perficit et infra ipsam sphaeram tertiam mole adhuc densiorem congregationr reliquit. Atque ad hunc ordinem processit ipsa congregatio disgregans, donec complerentur novem sphaerae caelestes et congregaretur inter sphaeram nonam infimam moles densata, quae esset quattuor elementorum materia* (Eng. trans. by Clare C. Riedl, op. cit., p. 13).

are completely different from those he says when wears the habit of the theologian. In contrast to the usual case, in this case "the habit makes the monk".

Our witty remark aside, the author of *On Light* says that the firmament (the first body formed in the outermost part of the sphere) diffuses its light from every part of itself to the center of the universe (*ab omni parte sua in centrum totius*; the genitive *totius* is translated by all as "*of the universe*"). Therefore one is led to think that light, by spherically expanding from its origin, has created in the outermost sphere the firmament (which one imagines as a spherical shell) and from the surface of it the light expands inwards (since he says "*in centrum totius*") and creates by subsequent rarefactions and condensations the nine heavenly spheres and the four elements. And all this has nothing to do with the cosmogony represented in Genesis.

Therefore it seems that Grosseteste, as exegete of Genesis, limits himself to reviewing the work of the exegetes who preceded him, although in some cases he expressing his opinions about what he is reporting. As a scientist, he feels free to present a theory of the origin of the universe (an Aristotelian universe) in which the divine intervention remains in the background.

2.2.8 St. Thomas Aquinas (1225–1274)

From the Franciscan Grosseteste we pass now to the Dominican Thomas Aquinas. We point this out to remind the reader that we are always moving among "philosophers" belonging to religious orders and therefore, generally, devoted to theological studies.

Thomas was born at Roccasecca (in present-day Lazio) almost assuredly in 1225 and began his early education in the Benedictine abbey of Monte Cassino, and later continued at the University of Naples (1239), where he came to know the Order of the friar preachers, recently founded by St. Dominic. He entered the Dominican Order in 1244. Overcoming the opposition of his family, he pursued his studies first in Paris and then in Cologne, where he was a student of Albert the Great. He went back to Paris as a teacher, both a first time and a second time, separated by a period (1259–1269) spent in Italy following the papal court. After the second Parisian period, he came back in Italy in 1272 and died there, still relatively young, in 1274, during a journey to the Council of Lyon.

We have recalled these dates to point out the contemporaneity of Thomas with the other philosophers of whom we have spoken until now. Obviously, it does not fall within our scope to speak about Thomas's philosophy and his boundless literary production in the theological-philosophical field. We limit ourselves to recalling that Thomas was, together with his master Albert the Great, the author of the great work of composition of Aristotelian philosophy with the Catholic doctrine. It is for this that he is become one of the foremost theological and philosophical pillars of the Catholic Church.

One of his most important works is the *Summa Theologica*,[109] which he began to write during his stay in Rome and was not able to complete before his death. It is in this work that the problem of the waters above the firmament is dealt with in detail. The *Summa* is divided into three parts. Each of the three parts is divided into treatises, each treatise is divided into questions (*quaestiones*),and each question is divided into articles (*articuli*). In order to facilitate the interpretation of the excerpts we shall report, we add that the article consists of four parts: a series of objections (introduced by *Videtur quod*); the statement of the author's thesis (introduced by *Sed contra*); the defense of this thesis (*Respondeo*); the solution of the objections (*Ad primum, secundum, etc.*).

---

[109] The Latin text is available online at http://www.thelatinlibrary.com/aquinas/summa.shtml (accessed 20.01.2020). The English translation by Alfred J. Freddoso is available online at:
https://www3.nd.edu/~afreddos/summa-translation/TOC.htm (accessed 20.01.2020).

In the proem of the work, Thomas points out that he intends to address all the faithful, beginners and advanced alike, in his explanation the Christian religion: "Since, according to the Apostle in 1 Corinthians 3:1–2 ('As unto little ones in Christ, I gave you milk to drink, not meat'), a teacher of Catholic truth not only ought to instruct those who are advanced, but is also charged with teaching beginners, our intention in the present work is to propound the things belonging to the Christian religion in a way consonant with the education of beginners".[110]

Remembering these words, in order to appreciate the clarity of the presentation, in the first part of the *Summa* we find the *Tractatus de Deo* in three parts, the third of which contains 77 questions of the total 119. Six of these concern the days of creation; the 68th relates to the second day and is divided into four articles, of which the two middle ones concern the existence of waters above the firmament and if the firmament separates the waters from the waters. We shall deal with these two articles. Let us start from the one which concerns the existence of the waters:

It seems that there are no waters above the firmament:

Objection 1: Water is naturally heavy. But the proper place of what is heavy is just down below and not up above. Therefore, there are no waters above the firmament.

Objection 2: Water is naturally a fluid. But, as is clear from experience, a fluid cannot stay still on the surface of a round body. Therefore, since the firmament is a round body, there cannot be water above the firmament.

Objection 3: Since water is an element, it is ordered toward the generation of mixed bodies, in the way that what is incomplete (*imperfectum*) is ordered toward what is complete (*perfectum*). But the place for such mixing is upon the earth and not above the firmament. Therefore, water would be useless above the firmament. But nothing in the works of God is useless. Therefore, there are no waters above the firmament.

But contrary to this: Genesis 1:7 says, 'He divided the waters that were above the firmament from those that were under the firmament'.

I respond: As Augustine says in *Super Genesim ad Litteram 2*, 'The authority of this passage of Scripture is greater than all the capability of human genius. Hence, whatever sort of waters these were, and in whatever way they were there, we do not at all doubt that they were there.'

However, the question of what sort of waters these are is not answered in the same way by everyone.

For Origen says that the waters above the heavens are the spiritual substances, and this is why Psalm 148:4 says, 'Let the waters that are above the heavens praise the name of the Lord,' and Daniel 3:60 says, 'Bless the Lord, all you waters that are above the heavens'.

But to this Basil replies in *Hexameron 3* that these things are said not because the waters are rational creatures, but because 'the consideration of them, contemplated thoughtfully by beings with understanding, brings to completion the glorification of their Creator'. Hence, in the same place (Daniel 3) the same thing

---

[110] Thomas of Acquinas, Summa, Prologus: *Quia Catholicae veritatis doctor non solum provectos debet instruere, sed ad eum pertinet etiam incipientes erudire, secundum illud apostoli I ad Corinth. III., tanquam parvulis in Christo, lac vobis potum dedi, non escam; propositum nostrae intentionis in hoc opere est, ea quae ad Christianam religionem pertinent, eo modo tradere, secundum quod congruit ad eruditionem incipientium* (Eng. trans. Alfred J. Freddoso, op. cit.).

is said about fire and hail and other things of this sort, which are clearly not rational creatures.

Therefore, one should claim that the waters are corporeal. However, it is necessary to specify in different ways what kind of waters they are, in accord with the different opinions about the firmament.

For if the firmament is understood to be the starry heaven and is claimed to be of the nature of the four elements, then by parity of reasoning the waters that are above the heaven can be believed to be of the same nature as elemental waters.

On the other hand, if the firmament is understood to be starry heaven but not of the nature of the four elements, then the waters above the firmament will not be of the nature of elementary waters. Instead, just as, according to Strabo, one heaven is called 'empyrean', i.e., fiery, only because of its splendorous light (*splendor*), so too the other heaven, which is above the starry heaven, will be called 'aqueous' (*aqueum*) only because of its transparency.

Again, if one claims that the firmament has a nature different from that of the four elements, then, as Augustine points out in *Super Genesim contra Manichaeos,* he can still say that the firmament divides the waters if we mean by 'water' the unformed matter of bodies and not the element water. For on this view, whatever lies between bodies divides waters from waters.

However, if the firmament is understood to be that part of the air in which clouds gather, then the waters that are above the firmament are those waters which, having been evaporated (*vaporabiliter resolutae*), are elevated above the part of the air from which rain is generated. For it is altogether impossible to claim, as some have (Augustine touches on this view in *Super Genesim ad Litteram* 2), that evaporated waters are elevated above the starry heaven—and this (a) because of the solidity of the heavens, (b) because of the intermediate region of fire, which would consume vapors of this sort, c) because the place where light and rarefied things are carried is under the curve of the moon's orbit, and also (d) because vapors do not appear to the senses to be elevated even as far as the peaks of certain mountains. What's more, the reply that the rarefaction of a body goes on *ad infinitum* because bodies are infinitely divisible is groundless. For a natural body is divided or rarified only to a set limit and not *ad infinitum*.

Reply to objection 1: To some it seems that the correct response to this argument is that even though the waters are naturally heavy, they are retained above the heavens by God's power. But in *Super Genesim ad Litteram 2* Augustine rules out this response, saying, 'At this point it is fitting to inquire into how God made the natures of things, and not what He intended to do in them by His miraculous power.'

Hence, one should respond alternatively that, given the last two opinions about the waters and the firmament, the solution is clear from what has been said. According to the first opinion, one has to posit an order among the elements that is different from the one Aristotle posits, so that certain dense waters surround the earth, whereas certain rarified waters surround the heavens—with the result that those waters are related to the heavens in the same way that these waters here below are related to the earth. Yet another response, as has been explained, is that 'water' here means the matter of bodies.

Reply to objection 2: Given the last two opinions mentioned above, the answer here is clear from what was said above.

On the other hand, given the first opinion mentioned above, Basil has two replies. The first is that it is not necessary that everything that appears round on its concave side is also round up above on its convex side. Second, the waters that are above the heavens are not fluids, but are, as it were, firmed up with a glacier-like solidity around the heavens. That is why some call these waters the 'crystalline' heaven.

Reply to objection 3: According to the third opinion, the waters above the firmament are elevated as vapor because of the usefulness of rain.

By contrast, given the second opinion, the waters are above the firmament, i.e., above the whole diaphanous heaven without stars. Some say that this heaven is the first moveable thing and that it turns all of heaven with the diurnal motion in order to effect, through the diurnal motion, the continuity of generation—just as the heaven in which the stars exist, through a motion that is in accord with the zodiac, effects the diversity of generation and corruption by approaching and receding and by the diverse powers of the stars.

On the other hand, given the first opinion, the waters are there, as Basil says, to temper the heat of the celestial bodies. As Augustine points out, some take as an indication of this the fact that the star Saturn is the coldest because of the proximity to the higher waters.[111]

---

[111] Thomas of Aquinas, *Summa*, Pt. I, Question 68, Art. 2: *Videtur quod acquae non sunt supra firmamentum. Aqua enim est naturaliter gravis. Locus autem propius gravis non est esse sursum, sed solum deorsum. Ergo aquae non sunt supra firmamentum.*

Praeterea, naturaliter aqua est fluida. Sed quod est fluidum, non potest consistere super corpus rotundum, ut experimentum patet. Ergo, cum firmamentum sit corpus rotundum, aqua non potest esse supra firmamentum. Praeterea, aqua, cum sit elementum, ordinatur ad generationem corporis mixti; sicut imperfectum ordinatur ad perfectum. Sed supra firmamentum non est locus mixtionis, sed supra terram. Ergo frustra aqua esset supra firmamentum. Nihil autem in operibus Dei est frustra. Ergo aquae non sunt super firmamentum.

Sed contra est quod dicitur Gen. I, quod divisit aquas quae erant supra firmamentum, ab his quae erant sub firmamento.

Respondeo dicendum quod, sicut dicit Augustinus, Ii super Gen. ad Litt. maior est Scripturae huius auctoritas quam omnis humani ingenii capacitas. Unde quomodo et quales aquae ibi sint, eas tamen ibi esse, minime dubitamus. *Quales autem sint illae aquae, non eodem modo ab omnibus assignatur. Origenes enim dicit quod aquae illae quae super caelos sunt, sunt spitituales substantiae, unde in Psalmo CXLVIII, dicitur, aquae quae super caelos sunt, laudent nomen domini; et Dan. III, benedicite, aquae omnes quae super caelos sunt, domino. Sed ad hoc respondet Basilius, in III Hexaem., quod hoc non dicitur eo quod aquae sint rationales creaturae; sed quia* consideratio earum, prudenter a sensum habentibus contemplata, glorificationem perficit creatoris. *Unde ibidem dicitur idem de igne et grandine et huiusmodi, de quibus constat quod non sunt rationales creaturae. Dicendum est ergo quod sunt aquae corporales. Sed quales aquae sint, oportet dversimode definire, secundum diversam de firmamento sententiam. Si enim per firmamentum intelligitur caelum sidereum quod ponitur esse de natura quatuor elementorum, pari ratione et aquae quae super caelos sunt, eiusdem naturae poterunt credi cum elementaribus aquis. Si autem per firmamentum intelligatur caelum sidereum quod non sit de natura quatuor elementorum, tunc et aquae illae quae sunt supra firmamentum, non erunt de natura elementarium aquarum, sed sicut, secundum Strabum, dicitur caelum empyreum, idest igneum, propter solum splendorem; ita dicetur aliud caelum aqueum propter solam diaphaneitatem, quod est supra caelum sidereum. Posito etiam quod firmamentum sit alterius naturae praeter quatuor elementa, adhuc potest dici quod aquas dividit, si per aquam non elementum aquae, sed materiam informem corporum intelligamus, ut Augustinus dicit, super Gen. contra Manich., quia secundum hoc, quidquid est inter corpora, dividit aquas ab aquis. Si autem per firmamentum intelligatur pars aeris in qua nubes condensantur, sic aquae quae supra firmamentum sunt, sunt aquae quae, vaporabiliter resolutae, supra aliquam partem aeris elevantur ex quibus pluviae generantur. Dicere enim quod aquae vaporabiliter resolutae eleventur supra caelum sidereum, ut quidam dixerunt, quorum opinionem Augustinus tangit in II super Gen. ad Litt. est omnino impossibile. Tum propter soliditatem caeli. Tum propter regionem ignis mediam, quae huiusmodi vapores consumeret. Tum quia locus quo feruntur levia et rara, est infra concavum orbis lunae. Tum etiam quia sensibiliter apparet vapores non elevari usque ad cacumina quorundam montium. Quod*

At last, let us pass to the article about the division of the waters:

It seems that the firmament does not divide waters from waters:

Objection 1: There is just one natural place for one body according to its species. But as the Philosopher says, all water is the same in species as all other water. Therefore, waters are not distinct from waters with respect to place.

Objection 2: Someone might reply that the waters above the firmament differ in species from the waters below the firmament.

Against this: Things that are diverse in species do not need anything else to distinguish them. Therefore, if the higher waters and lower waters differ in species, then it is not the firmament that distinguishes them from one another.

Objection 3: It seems that what divides waters from waters is something touched by waters on both sides, like a wall built in the middle of a river. But it is clear that the lower waters do not reach all the way to the firmament. Therefore, the firmament does not divide waters from waters.

But contrary to this: Genesis 1:6 says, 'Let there be a firmament made in the middle of the waters, and let it divide waters from waters'.

I respond: If one looked at just the surface of the text of Genesis, he could construct a picture that corresponds to the positions of certain ancient philosophers. For some of them claimed that water is an infinite body and the principle of all other bodies. Indeed, one could read the immensity of the waters into the name 'deep' (*abyssum*), when it says, 'Darkness was upon the face of the deep.' In addition, they claimed that the sensible heaven which we see does not contain all corporeal things under itself, but that there is an infinite body of waters above the heaven. And so one could claim that the firmament of the heaven divides the outer waters from the inner waters, i.e., from all the bodies contained below the heaven whose principle they claimed to be water.

However, since this position is shown to be false by sound arguments, one should not claim that this is what Scripture means.

---

etiam dicunt de rarefactione corporis in infinitum, propter hoc quod corpus est in infinitum divisibile, vanum est. Non enim corpus naturale in infinitum dividitur aut rarefit, sed usque ad certum terminum.

Ad primum ergo dicendum quod quibusdam videtur ratio illa solvenda per hoc, quod aquae, quamvis sint naturaliter graves, virtute tamen divina super caelos continentur. Sed hanc solutionem Augustinus excludit, II Lib. super Gen. ad Litt., dicens quod nunc quemadmodum Deus instituit naturas rerum convenit quaerere; non quid in eis ad miraculum suae potentiae velit operari. *Unde aliter dicendum est quod, secundum duas ultimas opiniones de aquis et firmamento, patet solutio ex praemissis. Secundum autem primam opinionem, oportet ponere alium ordinem in elementis quam Aristoteles ponat; ut quaedam aquae spissae sunt circa terram, quaedam vero tenues circa caelum; ut sic se habeant illae ad caelum, sicut istae ad terram. Vel quod per aquam intelligatur materia corporum, ut dictum est.*

Ad secundum etiam patet solutio ex praemissis, secundum duas ultimas opiniones. Secundum vero primam, respondet Basilius dupliciter. Unomodo, quia non est necessarium ut omne quod in concavo apparet rotundum, sit etiam supra rotundum secundum convexum. Secundo, quia aquae quae sunt supra caelos, non sunt fluidae; sed quasi glaciali soliditate circa caelum firmatae. Unde et a quibusdam dicuntur caelum crystallinum.

Ad tertium dicendum quod, secundum tertiam opinionem, aquae sunt supra firmamentum vaporabiliter elevatae propter utilitatem pluviarum. Secundum vero secundam opinionem, aquae sunt supra firmamentum, idest caelum totum diaphanum absque stellis. Quod quidam ponunt primum mobile, quod revolvit totum caelum motu diurno, ut operetur per motum diurnum continuitatem generationis, sicut caelum in quo sunt sidera, per motum qui est secundum zodiacum, operatur diversitatem generationis et corruptionis, per accessum et recessum, et per diversas virtutes stellarum. Secundum vero primam opinionem, aquae sunt ibi, ut Basilius dicit, ad contemperandum calorem caelestium corporum. Cuius signum acceperunt aliqui, ut Augustinus dicit, quod stella Saturni, propter vicinitatem aquarum superiorum, est frigidissima (Eng. trans. Alfred J. Freddoso, op. cit.).

Instead, consider that Moses was speaking to an uneducated people and, in accommodating himself to their intellectual weakness, he proposed to them only what is manifestly obvious to the senses. Now everyone, no matter how uneducated, perceives through the senses that earth and water are bodies. Air, however, is not perceived by everyone to be a body, since even some philosophers haves claimed that air is nothingness, calling what is full of air a vacuum. And so Moses explicitly mentions water and earth, but does not explicitly name air, so as not to propose something unknown to the uneducated. Yet in order to express the truth to those capable of understanding it, he makes room for an interpretation involving air by signifying it as connected to the water, when he says: 'Darkness was upon the face of the deep'—which means that a diaphanous body, the subject of light and darkness, is upon the face of the water. So, then, regardless of whether we mean by the firmament the starry heaven or the cloudy part of the air, it is appropriate to say that the firmament divides waters from waters, given either that 'water' means unformed matter or that all diaphanous bodies are understood by the name 'waters'. For the starry heaven divides the lower diaphanous bodies from the higher bodies, whereas the cloudy air divides the higher part of the air, in which rains and similar irruptions are generated, from the lower part of the air, which is connected with water and is understood under the name 'waters'.

Reply to objection 1: If 'firmament' means the starry heaven, then the higher waters are not the same in species as the lower waters.

On the other hand, if 'firmament' means the cloudy air, then both waters are of the same species. And in that case the two places are not assigned to the waters for the same reason; instead, the higher place is the place of the generation of waters, whereas the lower place is the place of rest for those waters.

Reply to objection 2: If the waters are taken to be diverse in species, the firmament is said to divide waters from waters not in the sense that it causes the distinction between them, but in the sense that it is the terminus of both sorts of waters.

Reply to objection 3: Because of the invisibility of the air and of similar bodies, Moses includes all bodies of this sort under the name 'waters. And in this way it is clear that there are waters on both sides of the firmament, no matter how 'firmament' is understood.[112]

---

[112] Thomas of Aquinas, *Summa*, Pt. I, Question 68, Art. 3: *Videtur quod firmamentum non dividat aquas ab aquis. Unius enim corporis secundum speciem, est unus locus naturalis. Sed omnis aqua omni aquae est eadem specie, ut dicit philosophus. Non ergo aquae ab aquis sunt distinguendae secundum locum. Si dicatur quod aquae illae quae sunt supra firmamentum, sunt alterius speciei ab aquis quae sunt sub firmamento, contra, ea quae sunt secundum speciem diversa, non indigent aliquo alio distinguente. Si ergo aquae superiores et inferiores specie differunt, firmamentum eas ab invicem non distinguit.*
Praeterea, illud videtur aquas ab aquis distinguere, quod ex utraque parte ab aquis contingitur; sicut si aliquis paries fabricetur in medio fluminis. Manifestum est autem quod aquae inferiores non pertingunt usque ad firmamentum. Ergo non dividit firmamentum aquas ab aquis.
Sed contra est quod dicitur Gen. I, fiat firmamentum in medio aquarum, et dividat aquas ab aquis.
Respondeo dicendum quod aliquis, considerando superficie tenus litteram Genesis, posset talem imaginationem concipere, secundum quorundam antiquorum philosophorum positionem. Posuerunt enim quidam aquam esse quoddam infinitum corpus, et omnium aliorum corporum principium. Quam quidem immensitatem aquarum accipere posset in nomine abyssi, cum dicitur quod tenebrae erant super faciem abyssi. Ponebant etiam quod istud caelum sensibile quod videmus, non continet infra se omnia corporalia; sed est infinitum aquarum corpus supra caelum. Et ita posset aliquis dicere quod firmamentum caeli dividit aquas exteriores ab aquis interioribus, idest ab omnibus corporibus quae infra caelum continentur, quorum principium aquam ponebant. Sed quia ista positio

As already mentioned, Thomas writes his work with the aim of teaching, that is, of explaining the Christian doctrine to all, including beginners (*etiam incipientes erudire*). Since this can never occur directly (the *incipientes* never will be able to read the *Summa*), Thomas studies all *ad litteram* interpretations given by the exegetes who have preceded him and passes along those he considers defensible and, therefore, that the preachers will be allowed to tell. As one can see, Bede's "congealed waters" fails the examination. In any case, we can say that Thomas makes a screening of the interpretations he considers plausible, but does not suggest any new interpretation.

2.2.9 Nicholas of Cusa (1401–1464)

With Thomas Aquinas the cycle of Scholastic philosophy comes to an end, and as far as we are concerned, we can consider his treatment of the works of the second day of Genesis essentially as a conclusion of the medieval commentaries on that subject. We can, in a certain sense, maintain that the Middle Ages have no more new elements to supply us in the field of commentaries on Genesis. But there is a remarkable improvement in the approach to the problem of the creation and this improvement is made in a work (we shall deal with it below) by the figure who is considered (at the present time) to be the most important philosopher of the fifteenth century: Nicholas of Cusa (Nicolaus Cusanus). I say parenthetically "*at the present time*" since substantially Cusanus has been *rediscovered* in the twentieth century. In fact, it was in 1927 that the Neo-Kantian philosopher Ernst Cassirer[113] identified Cusanus as the initiator of modern philosophy, by finding in his work anticipations of modern themes and, in the specific case of natural philosophy, anticipations of Copernicus, Kepler and Galileo.

Cusanus was a known and respected scholar in the world of the Church and among the scholars of the ancient books and manuscripts, but not for his philosophical works. Before

---

per veras rationes falsa deprehenditur, non est dicendum hunc esse intellectum Scripturae. Sed considerandum est quod Moyses rudi populo loquebatur, quorum imbecillitati condescendens, illa solum eis proposuit, quae manifeste sensui apparent. Omnes autem, quantumcunque rudes, terram et aquam esse corpora sensu deprehendunt. Aer autem non percipitur ab omnibus esse corpus, intantum quod etiam quidam philosophi aerem dixerunt nihil esse, plenum aere vacuum nominantes. Et ideo Moyses de aqua et terra mentionem facit expressam, aerem autem non expresse nominat, ne rudibus quoddam ignotum proponeret. Ut tamen capacibus veritatem exprimeret, dat locum intelligendi aerem, significans ipsum quasi aquae annexum, cum dicit quod tenebras erant super faciem abyssi; per quod datur intelligi super faciem aquae esse aliquod corpus diaphanum quod est subiectum lucis et tenebrarum. Sic igitur sive per firmamentum intelligamus caelum in quo sunt sidera, sive spatium aeris nubilosum, convenienter dicitur quod firmamentum dividit aquas ab aquis, secundum quod per aquam materia informis significatur; vel secundum quod omnia corpora diaphana sub nomine aquarum intelliguntur. Nam caelum sidereum distinguit corpora inferiora diaphana a superioribus, Aer vero nubilosus distinguit superioris aeris partem, in qua generantur pluviae et huiusmodi impressiones, ab inferiori parte aeris, quae aquae connectitur, et sub nomine aquarum intelligitur.

Ad primum ergo dicendum quod, si per firmamentum intelligatur caelum sidereum, aquae superiores non sunt eiusdem speciei cum inferioribus. Si autem per firmamentum intelligatur nubilosus aer, tunc utraeque aquae sunt eiusdem speciei. Et deputantur tunc duo loca aquis non eadem ratione; sed locus superior est locus generationis aquarum, locus autem inferior est locus quietis earum.

Ad secundum dicendum quod, si accipiantur aquae diversae secundum speciem, firmamentum dicitur dividere aquas ab aquis, non sicut causa faciens divisionem; sed sicut terminus utrarumque aquarum.

Ad tertium dicendum quod Moyses, propter invisibilitatem aeris et similium corporum, omnia huiusmodi corpora sub aquae nomine comprehendit. Et sic manifestum est quod ex utraque parte firmamenti, qualitercumque accepti, sunt aquae (Eng. trans. Alfred J. Freddoso, op. cit.).

[113]Ernst Cassirer, *Individuum und Cosmos in der Philosophie der Renaissance* (Leipzig: G. B. Teubner, 1927). See the English translation: *The Individual and the Cosmos in the Renaissance Philosophy*, trans. Mario Domandi (The University of Chicago Press, 1963).

examining the excerpts from his work we are interested in, let us supply a brief biographical sketch.

Nicholas Krebs (or Chryffz) was born in 1401 in Kues (in Latin Cusa, hence the name Cusanus) near Trier (Germany). As a child he studied in Deventer at the Latin school of the Brothers of the Common Life. He was only 15 when he enrolled in the University of Heildelberg and then, from 1418 to 1423, in Padua, where he received the degree of Doctor of Laws. In Padua, besides canon law, he also studied mathematics and astronomy. There, he met the cardinal Giuliano Cesarini, who had a great influence in the career of Cusanus in the world of the Church. Over the course of his life Cusanus held very important positions (he also attended, in 1432, the Council of Basel, during which the reunification of the Latin Church with the Greek Church was to be decided). In 1450 he was also named, by Pope Nicholas V, prince-bishop of Brixen (Bressanone), but he had difficulties in holding the position since the Tyroleans did not accept the pope's decision and their duke Sigismund was always opposed to the appointment of Cusanus. In the final years of his life, Cusanus was cardinal in Rome, where he arrived in 1461, and there was seriously ill for a long time. He died in Todi on 11 August 1464. Cusanus, besides having been a man of deep studies and high culture, performed many important tasks as an envoy of the pope. It does not rest with us to give an exposition, even if concise, of Cusanus's philosophy, that is, of *docta ignorantia*, or "learned ignorance".[114] According to the method we have followed up to now, we shall limit ourselves to singling out in his work the loci which concern the problem we are dealing with.

As we have already said, Cusanus wrote no commentaries on Genesis, but did instead write a dialogue (*Dialogus de Genesi*), in which the problem of creation is treated within the philosophy of learned ignorance.[115] The dialogue was written in 1447 and in it the interlocutor is a certain Conrad (probably Conrad of Wartberg, canon of the monastery of Münster-Meinfeld). It starts with a philosophical consideration on God meant as "*idem absolutum*" and then addresses the question of the reality of Genesis. After having expounded his philosophical explanation of creation, Cusanus deals with the problem of its compatibility with the biblical narration of Genesis, establishing to that end some guidelines for interpreting Scripture. On this, we report the following excerpt: "But where Moses expressed in human terms the manner in which all these things were done, I believe that he elegantly expressed it to the end [of saying] what is true in the manner in which what is true can be grasped by man. But you know that he used a human manner in order to instruct men in human terms. To these things he added, in their own place and subsequently to the human manner-of-speaking, other things of such kind that intelligent men would understand that the things which express the manner [of creating] are human assimilations for the unattainable divine manner. For when Moses revealed that God is nothing of all the things that can be seen or depicted or carved and that He is visible by man only in vestiges that are subsequent to Him and that He who is of infinite power does nothing through temporal delays, Moses showed sufficiently that he refigured in human terms the mode of the inexpressible act of creation. Hence, the wise, who say that the invisible God created *at once* all things as he willed to, do not contradict the intent of the law-giver Moses, even as the

---

[114]To this end, we indicate to the reader the Seminar Notes (Yale University, 2015) by Karsten Harries: Nicholas of Cusa-*On Learned Ignorance*, available online at https://cpb-us-w2.wpmucdn.com/campuspress.yale.edu/dist/8/1250/files/2012/09/Cusanus-On-Learned-Igorance-17z8dxd.pdf (accessed 20.01.2020).

[115] The Latin text is from *Nicolai de Cusa Opera omnia iussu et auctoritate Academiae litterarum heidelbergensis ad codicum fidem edita*, vols, 4-5, eds. Ernst Hoffmann and Raymond Klibansky (Hamburg: Felix Meiner, 1959). For the English translation: Jasper Hopkins, *Nicholas of Cusa On Learned Ignorance. A Translation and an Appraisal of De Docta Ignorantia* (Minneapolis: Arthur J. Banning Press, 1985).

very many others also do not do who have used other refiguring modes. And with this befigurement is especially compatible the fact that when Moses spoke of man, he called him *Adam*, [a word] which is an appellative enfolding in its signification *man*, whether masculine or feminine.

"And for the aforementioned reasons, and for many others that can be treated more suitably elsewhere, the Jews are enjoined to reserve the beginning of Genesis for the wise, in order that the literal, surface-meaning would not offend neophytes. But the wise and those who are quite skilled in theological matters, knowing that the divine modes cannot be apprehended, are not offended if the refiguring assimilation is found to be contracted to the language of the heavens. For, as best they can, the wise free it from the contraction in order to see that only the Absolute Same causes-to-be-identical. Hence, the discrepancies (a) between the historical accounts and (b) between rationales, times, names, and men, and the inaccurate account of the flow of the rivers that are said to flow from the midst of Paradise, and whatever other [discrepancies], even were they more absurd, do not all offend them. Rather, from among the absurdities they seek out the more secret mysteries—just as in the case of the especially accomplished intellects of the saints you can also find regarding that [first] part of Genesis, if you read Ambrose's *De Paradiso* and his *In Hexameron* and if you read Basil, Augustine, Jerome, and the like. I noticed that, being wise, all these [saints], although they seem to be odds in many respects, agree in the main. Still, not all take literally the manner [of creating] that is narrated there. The views of all these [men] concerning the manner [of God's creating] I accept in the following way: viz., as if they were wise men's different concepts of [that] inexpressible manner; and turning myself only to the Same—which each [of those wise] has endeavored to refigure assimilatively—I find rest in it."[116]

Therefore, Cusanus maintains that the narration of Genesis should not be understood "*ad litteram*", or at least not always, but must be interpreted. Like others earlier, including, as we know, St. Thomas, Cusanus says that Moses wrote in a manner that men could understand and

---

[116] Nicholas of Cusa, *Dialogus de Genesi* II, 159, 160: *Ubi vero Moyses modum, quo haec acta sunt omnia, humaniter exprimit, credo ipsum ad finem, ut verum modo quo verum per hominem capi posset, eleganter expressisse. Sed usum scis modo humano ad finem, ut homines humaniter instruat, quibus post humanum modum adicit suo loco talia, ut intelligentes intelligant illa, quae modum exprimunt, inattingibilis divini modi fore humanam assimilationem. Nam quando aperuit deum nihil omnium esse, quae videri aut figurari aut insculpi possunt, atque quod ipse solum in vestigiis, quae sunt posteriora eius, visibilis sit per hominem, quodque ipse infinitae potentiae nihil agat per temporales moras, satis ostendit se creationis inexpressibilis modum humaniter configurasse.*

Unde sapientes, qui invisibilem deum omnia, ut voluit, simul aiunt creasse, non contradicunt intentioni legislatoris Moysi, sicut nec alii plerique, qui alios confinxerunt modos. Et ad hoc maxime facit, quia, cum de homine loqueretur, ipsum Adam appellat, quod est appellativum in suo significato hominem sive masculum sive feminam complicans.

Et ob praemissa atque alia multa, quae convenientius alibi tractari possunt, principium Geneseos prudentibus mandatur servari per Iudaeos, ne literalis superficies novicios offendat. Prudentes autem atque in theologicis peritiores scientes divinos modos sine apprehensibili modo esse non offenduntur, si configuralis assimilatorius ad consuetudinem audientium contractus reperitur. Ipsi enim absolvunt eum a contractione illa, quantum eis possibile fuerit, ut intueantur tantum idem absolutum identificare. Hinc eosdem nec diversitas historiarum, rationum, temporis, nominum, hominum, adversitas fluxus fluviorum, qui ex medio paradisi narrantur effluere, et quaeque alia, etiamsi forent absurdiora, minime offendunt, sed mysteria secretiora ex absurdioribus venantur, sicut in exercitatis maximis ingeniis sanctorum circa eam Geneseos partem reperire poteris, si Ambrosium De Paradiso et eundem in Hexameron, Basilium, Augustinum, Hieronymum et tales lectitaveris. Quales omnes, licet discrepare in plerisque videantur, adverti uti prudentes in principali concurrere, licet modum non omnes ad litteram admittant ibidem narratum. Quorum omnium considerationem circa modum sic accepto, quasi sint sapientum varii conceptus inexpressibilis modi, non nisi me ad idem ipsum, quod quisque nisus est assimilatorie configurare, convertens et in eo quiescens (Eng. trans. Jasper Hopkins, op. cit., pp. 402-403).

thus he chose a language suitable to that end. To strengthen this claim, he also quotes the historical knowledge (which we have already come across) that the Jews reserved the beginning of Genesis to the wise, since the literal sense might have offended the neophytes. Substantially, Cusanus frees himself from the attitude of the preceding medieval annotators, and is always careful in reading and commenting all details of the narration.

As we have recalled above, the *Dialogus de Genesi* dates to 1447, whilst his main work, *De docta ignorantia*, dates to about ten years before. *De docta ignorantia*, together with *De ludo globi* (1462), is often quoted, in addition to its philosophical relevance, for the "anticipations" of the theories of Copernicus and Galileo. Whereas in *De ludo globi*, it is Galileo's principle of inertia to be "anticipated",[117] in *De docta ignorantia*, besides attention being drawn to the relativity of motion in general (see Book II, chap XI), the motion of the earth is also dealt with (Book II, Chap. XII). We cannot renounce quoting the beginning of the Chap. XI: "Perhaps those who will read the following previously unheard of [doctrines] will be amazed, since learned ignorance shows these [doctrines] to be true".[118] Among the unheard things, as we have said, there is the motion of the earth and, above all, the denial of its central position. One might say that the centrality of the earth is made relative. In this regard, Harries wonders, "But is this really the position of Cusanus? At this point it is not yet altogether clear what he is asserting. Does it really entail a radical break with the Aristotelian-Ptolemaic understanding of the cosmos or does it just introduce into it something like an uncertainty principle?".[119]

In any case, Cusanus writes: "Therefore, the earth, which cannot be the center, cannot be devoid of all motion. Indeed, it is even necessary that the earth be moved in such a way that it could be moved infinitely less. Therefore, just as the earth is not the center of the world, so the sphere of fixed stars is not its circumference—although when we compare the earth with the sky, the former seems [*videatur*] to be nearer to the center, and the latter nearer to the circumference. Therefore, the earth is not the center either of the eighth sphere or of any other sphere".[120]

After having again spoken about the universe and asserted that its center is God, he says: "The ancients did not attain unto the points already made, for they lacked learned ignorance. It has already become evident to us that the earth is indeed moved, even though we do not perceive this to be the case".[121]

And at the end, on the sphericity of the earth, he writes, "Moreover, the earth is not spherical, as some have said; yet, it tends toward sphericity, for the shape of the world is contracted in the world's parts, just as is [the world's motion]. Now, when an infinite line is considered as contracted in such a way that, as contracted, it cannot be more perfect and more capable, it is [seen to be] circular; for in a circle the beginning coincides with the end. Therefore, the most

---

[117]In regard to this, see Dino Boccaletti, *Galileo and the Equations of Motion* (Springer, 2016), p. 96.

[118] Nicholas of Cusa, *De docta ignoranza* II, 11, 156: *Fortassis admirabuntur, qui ista prius inaudita legerint, postquam ea vera esse docta ignorantia ostendit* (Eng. trans. Jasper Hopkins, op. cit., p. 89). In our opinion, Hopkins' English translation does not sufficiently reproduce the triunphal tone of the original Latin.

[119] K. Harries, Nicholas of Cusa-*On Learned Ignorance*, op. cit., p. 120.

[120] Nicholas of Cusa, *De docta ignoranza* II, 11: *Terra igitur, quae centrum esse nequit, motu omni carere non potest. Nam eam moveri taliter etiam necesse est, quod per infinitum minus moveri posset. Sicut igitur terra non est centrum mundi, ita nec sphaera fixarum stellarum eius circumferentia, quamvis etiam, comparando terram ad caelum, ipsa videatur centro propinquior et caelum circumferentiae. Non est igitur centrum terra neque octavae aut alterius sphaerae* (Eng. trans. Jasper Hopkins, op. cit., p. 90).

[121] Nicholas of Cusa, *De docta ignoranza* II, 12: *Ad ista iam dicta veteres non attigerunt, quia in docta ignorantia defecerunt. Iam nobis manifestum est terram istam in veritate moveri, licet nobis hoc non appareat* (Eng. trans. Jasper Hopkins, op. cit., p. 92).

nearly perfect motion is circular; and the most nearly perfect corporeal shape is therefore spherical".[122]

Here, as Harries remarks, the old Platonic axiom remains as a regulative ideal. Perhaps, we can conclude that the "anticipations" which Cusanus introduces are less disruptive than his triumphal tone leads one to perceive.

---

[122] Nicholas of Cusa, *De docta ignoranza* II, 12: *Terra etiam ista non est sphaerica, ut quidam dixerunt, licet tendat ad sphaericitatem. Nam figura mundi contracta est in eius partibus, sicut est motus; quando autem linea infinita consideratur ut contracta taliter, quod ut contracta perfectior esse nequit atque capacior, tunc est circularis; nam ibi principium coincidit cum fine. Motus igitur perfectior est circularis, et figura corporalis perfectior ex hoc sphaerica* (Eng. trans. Jasper Hopkins, op. cit., p. 93).

Chapter 3
The Waters above the Firmament in the Commentaries on Genesis in the Sixteenth and Seventeenth Centuries

As Arnold Williams says, in his work devoted to an account of the commentaries on Genesis published between 1527 and 1633, "To the student of ideas and to the historian of culture a somewhat detailed examination of the position of Genesis in the intellectual pattern of the Renaissance is productive of a fuller understanding of that grand epoch. Genesis is the first book of the Scriptures. It not merely satisfied the curiosity universal in all ages for authentic history of the primitive state of man, but it provided also an inspired foundation for the understanding of the whole development of history".[1]

The subject of our inquiry is, at least apparently, by far more confined, but we must nevertheless expect new arguments and a different attitude with respect to the medieval commentaries. In our opinion, among the new circumstances which could have influenced, one way or another, the annotators, two are particularly important. The first is the Reformation, which had its origin with the famous episode of the posting of the ninety five theses of Martin Luther on the portal of the church of the castle of Wittenberg in 1517, with the consequent theological disputes also regarding the Bible.

The second, which concerns particularly the Catholic annotators of the Bible, is the following: in the Fourth Session of the Tridentine Council (8 April 1546), it was established that the Vulgate (i.e., St. Hieronymus's translation of the Bible) is the only version of the Bible to be trusted authentic and that the commentaries can be published only if authorized by the competent authority of the Church, under the threat of excommunication.

It is obvious that this will greatly restrict the freedom of interpretation of those who, whether they like it or not, are integrated in the structures of the Catholic Church. Having said that, we shall deal now with commentaries, both Catholic and Protestant, starting from the commentaries written by Martin Luther.

3.1 Martin Luther (1483–1546)

The events of Luther's life are largely known and therefore it may seem to be superfluous to tell them again. Nevertheless, in this inquiry, it is also important to keep sight of the biographies (and, above all, of the dates), because of the suitable connections and comparisons.

Luther was born in Eisleben (Saxony) in 1483. He began his studies in Mansfeld and continued them in Magdeburg and then at Eisenach. In 1501 he joined the University of Erfurt, where he earned the degree of master of arts, in 1505 he entered the Augustinian Monastery of Erfurt in order to study theology and, starting from 1513, he was professor of biblical exegesis. The posting of the ninety five theses on the portal of the church of the castle of Wittenberg occurred in 1517, and in 1518 he was declared a heretic by the Pope. We neglect the rest,

---

[1] Arnold Williams, *The Common Expositor: An account of the Commentaries on Genesis 1527–1633* (The University of Carolina Press, 1948), p. 3.

© The Editor(s) (if applicable) and The Author(s), under exclusive license to Springer Nature Switzerland AG 2020
D. Boccaletti, *The Waters Above the Firmament*,
https://doi.org/10.1007/978-3-030-44168-5_3

recalling only that in 1534 he finished his famous translation of the Bible into German. He died in 1546 in his hometown of Eisleben and was buried in the church of Wittenberg.

Let us deal now with the commentaries on Genesis written by Luther: there are two and they are of different natures.

The first, *In Genesin, Mosi Librum Sanctissimum*, was published in 1527.[2] We learn from the *Praefatio* that this work resulted from the translation into Latin of the *Declamationes*, regarding the book of Genesis that he presented to "his" people of Wittenberg in the German language (*….populo meo Wittenbergensi declamationibus vernaculis tractavi,……*). The purpose, because of the intended readership, was not that of a doctrinal treatment (*non ut novum aliquod doctrine, aut eruditionis ederem*), but rather that of disseminating the knowledge of Genesis. Therefore we must not expect learned disquisitions about our problem. In fact, he says: "This inferior water is air and water always in motion, as we see. Whereas we don't know what water is above the heavens, only God knows it, here we leave the place to the Holy Spirit, who is more learned and knows more things than us. I would expect it would be air, since it stands under the heaven. But he who created all things, and the heaven from the water, can even maintain the water above the heaven. The bright and solid heaven has already been made, and indeed already made is that which on the first day was called heaven and which rightly and with already the suitable name is called heaven, since shortly before the earth deep down was simply called void and empty".[3]

As one can see, when push comes to shove, he resorts to the principle of authority; that is, God can do all he wants, even the things we consider impossible. For the "people of Wittenberg", this explanation might have been enough.

The second commentary, *In Primum Librum Mose*, dates to 1544 (*Die Natali Christi, Anno MDXLIIII*, as he says in the *Praefatio*)[4], two years before his death, and is a certainly mature work with ambitions far exceeding those of the earlier work. In any case, at the beginning, he safeguards himself with regard to the difficulties of interpretation: "The first chapter of the divine Book before us, is written in words, indeed, the most simple; but it contains things the most important, and at the same time the most obscure. Hence, it was forbidden among the Jews (according to the authority of Jerome) that any one should read these things himself, or speak of them to others, until he had attained his thirtieth year".[5]

It is curious that Luther, when introducing the works of the second day, expresses his disappointment because Moses does not speak about the creation and fall of the angels. He says that Moses seems have forgotten it (*Moses sui oblitus*) and then adds that in the rest of the Old Testament the subject is never dealt with either (*nihil omnino extat in scriptura*). But this is not enough for him: "It is wonderful, therefore, that Moses is wholly silent on things so great, and

---

[2] Martin Luther, *In Genesin, Mosi Librum Sanctissimum, D. Martini Lutheri Declamationes Praeterea Index, paucis opusculi totius summam continens* (Haganoae per Ioan. Secerium, (August) 1527).

[3] Martin Luther, *In Genesin,* op. cit., p. 7: *Haec aqua inferior, est aer et aqua inquieta, ut videmus. Qualis autem aqua sit supra coelos, nos nescimus, Solus Deus novit, nos hic locum relinquamus Spiritui Sancto, ut doctior sit et plus sciat quam nos. Ego suspicarer esse aerem, sicut adhuc est infra coelum. Sed qui creavit omnia, et coelum ex aqua, potest supra coelum aquam continere. Coelum iam lucidum et firmum factum est, et iam vere coelum factum est, quod primo die appellabatur coelum, iam merito et iusto nomine coelum vocatur, quod paulo ante terra simpliciter, imo inanis et vacua appellabatur* (our Eng. trans.).

[4] Martin Luther, *In Primum Librum Mose Enarrationes Reverendi Patris D.D. Martini Lutheri* (Wittenberg, 1544).

[5] Martin Luther, *In Primum Librum* , op. cit., I: *Primum caput simplicissimis quidem verbis est scriptum, sed res continet maximas et obscurissimas. Quare apud Ebreos (sicut D: Hieronymus testatur) prohibitum fuit, ne quis ante annum aetatis trigesimum illud legeret aut enarraret alijs* (Eng. trans. Henry Cole, from *Luther Still Speaking. The Creation: A Commentary on the first five chapters of the Book of Genesis by Martin Luther* (Edinburgh: T.& T. Clark, 1858), p. 23.

of such high interest. From this fact it has arisen, that men, having nothing certain recorded upon the deep subject, have naturally fallen into various fictions and fabrications: —that there were nine legions of angels; and that so vast was their multitude, that they were nine whole days falling from heaven. Others have indulged imaginations concerning the mighty battle between these superior beings: —in what manner the good, resisted the evil, angels. My belief is, that these ideas of the particulars of this battle, were taken from the fight which exists in the church; where godly ministers, are ever contending against evil and fanatical teachers".[6]

He then goes on to speak about the angels, concluding at the end that perhaps Moses neglected that subject since he had to speak to a new and uninformed people (*rudi et novo populo*) and for this spoke only about the subjects he believed necessary and useful for the people knew. We have already met with an argument of this kind on the part of Thomas Aquinas, and the same argument was also shared by Nicholas of Cusa.[7]

At last he comes to speak of the work of the second day: "We come now to the work of the *second day*, wherein we shall see, in what manner God produced, out of this original rough undigested mist, or nebulosity, which He called 'heaven', that glorious and beauteous 'heaven' *which* now is, and *as* it now is; if you except the stars, and the greater luminaries".[8]

At this point, Luther begins to detach himself from the text of the Vulgate. In fact, he begins from the Hebrew version of the Bible to substantiate the origin of the heaven from the water and again from the Hebraic text he takes the idea of the ultimate heaven as the result of an expansion: "This unformed body of mist, or nebulosity, created out of nothing, on the *first day*, God grasps, by His Word; and commands it to extend itself into the form, and with the motion, of a sphere. For in the Hebrew, the word RAKIA, signifies "a something extended"; from the verb RAKA; which signifies 'to unfold or expand'. And the heaven was formed by an *extension* of that original rude body of mist: just as the bladder of a hog is extended into a circular form, when it is inflated. I use thus a rustic similitude, that the sacred matter may be the more plainly understood".[9]

Because of its prolixity, we cannot report Luther's commentary in its integrity. He often departs from the subject, with quotations and argumentations in which he makes an exhibition of his biblical erudition, not always necessary to substantiate the context. Thus we must limit ourselves to the essential points, which still are considerably prolix. Hence let us go on with the nature of the heaven: "And the heaven is still more subtile and thinner than the air, or atmosphere. For its blue or sea-colour, or water-colour, appearance, is not a proof of its density, but rather of its distance, and its thinness: to which its rarefied state, if you compare the thicker substance of the clouds, the latter will appear, in the comparison, like the smoke of wet wood, when first ignited ... Wherefore the heaven which cannot consist by any boundary of its own,

---

[6] Martin Luther, *In Primum Librum...*, op. cit., p. VII: *Mirum igitur est tacere de his tantis rebus Mosen. Hinc factum est cum nihil certi haberent homines, ut aliquid fingerent, nempe, quod novem fuerint Angelorum chori, ac tanta multitudo, ut novem totos dies ceciderint. Finxerunt etiam de pugna maxima, qomodo boni Angeli restiterint malis. Hoc puto sumptum esse ex pugna Ecclesiae, quod sicut pij Doctores, pugnant contra malos et fanaticos* (Eng. trans. Henry Cole, op. cit., p. 45).

[7] See the excerpt in Chap. 2, footnote 116.

[8] Martin Luther, *In Primum Librum...*, op. cit., p. VII: *Hoc nunc secundi diei opus est, quomodo ex ista rudi nebula, quam coelum appellavit, produxerit coelum formosum et elegans, quale nunc est, si stellas et lumina maiora adimas* (Eng. trans. Henry Cole, op. cit., p. 46).

[9] Martin Luther, *In Primum Librum...*, op. cit., p. VII: *Istam rudem massam nebulae primo die ex nihilo conditae, apprehendit Deus per verbum, et iubet, ut extendatur in modum spherae. Nam vocabulum Rakia Ebraeis extensum quiddam significat a verbo Raka quod expandere et explicare significat. Nam coelum ita factum est, ut extenderetur ista massa rudis, sicut vesica suis circulari forma extenditur cum inflatur, utar enim rustica similitudine, quo res fiat planior* (Eng. trans. Henry Cole, op. cit., p. 47).

as being aqueous, consists by the Word of God; as we have it spoken in the present divine Record of Moses—'Let there be a firmament!'".[10]

And then: "This marvelous extension of the original rude and dense nebulosity, or cloud, or mist, is here called by Moses 'a firmament': in which, the sun, with all the planets, have their motion round the earth, in that most subtile material. But who is it that gives such *firmness* to this most volatile and fluctuating substance? Most certainly it is not nature that gives it: which, in far less important things than these, can exert no such power. It follows therefore, that it is the work of Him, who, 'in the beginning', said unto the heaven, and unto this volatile substance, 'Let there be a firmament'; or 'Be thou a firmament'; and who establishes and preserves all these things, by His omnipotent power, put forth through His Word. This Word makes the air, with all its thinness and lightness, to be harder and firmer than adamant, and to preserve its own boundary; and which Word could, on the contrary, make adamant to be softer than water: in order that, from such works as these, we might know what kind of a God our God is: namely, the God omnipotent; who made out of the rude mass of an unformed heaven, the present all-beauteous all-glorious heaven; and who did all these things, according to His will, as well as according to His power".[11]

But at this point Luther too, as did his predecessors before him, asks himself some questions: "It is a circumstance naturally exciting our particular wonder, that Moses evidently makes three distinct parts, or divisions, of this portion of the creation. He describes 'a firmament in the midst of the waters', which 'divides the waters from the waters'. —For myself, I am inclined to think, that the firmament here mentioned is the highest body of all; and that the waters (not those 'above' the firmament, but those which hang and fly about 'under' the firmament) are the clouds, which we behold with our natural eyes: so that, by the waters which are 'divided from the waters', we may understand the clouds, which are divided from our waters which are in the earth. Moses, however, speaks, in the plainest possible terms, both of waters 'above' and of the waters 'under' the firmament. Wherefore I here hold my own mind and judgment in captivity, and bow to the Word; although I cannot comprehend it.

"But a question here arises; —what those waters are, and how those bodies of water which are 'above' the firmament are distinguished from those which are 'under' the firmament. The division and distinction here made by philosophers is well known. They make the elements to be *four*: and they distinguish and place them according to their qualities. They assign the lowest place to the *earth*, a second place to the *water*; a third to the *air*; and the last and highest place to the *fire*. Other philosophers add to these *four* elements *aether*, as a *fifth* essence. After this division and number of the *elements*, there are enumerated *seven spheres* or *orbs* of the planets, and an *eighth sphere* of the fixed stars. And on these subjects it is agreed between all philosophers that there are *four spheres* of generating and corruptible principles; and also *eight*

---

[10] Martin Luther, *In Primum Librum…*, op. cit., p. VII: *At coelum etiam subtilius et tenuius est natura, quam iste aer. Nam quod apparet ceruleum esse, non est argumentum definitatis, sed longinquitatis et tenuitatis potius. Ad quam si nubium corpora conferas, sunt quasi ligni humidi fumus, quod accenditur … Igitur coelum quod suo termino non potest consistere, est enim aqueum, consistit verbo Dei, quod hic audimus, Fiat firmamentum* (Eng. trans. Henry Cole, op. cit., p. 48).

[11] Martin Luther, *In Primum Librum…*, op. cit., p. VIII: *Haec admirabilis istius crassae nebulae extensio. Firmamentum a Mose appellatur, in quo Sol cum reliquis Planetis suum motum habet circum terram, in illa subtilisima materia. Sed hanc firmitatem isti fluxili et vagae substantiae quis Autor addit: Profecto natura non facit, quae in levioribus id prestare non potest. Igitur istius hoc opus est, qui ad coelum et lubricam istam substantiam dicit: Sis Firmamentum, ac verbo ista omnia firmat et conservat pro sua omnipotentia. Hoc verbum facit, ut tenuissimus aer, sit durior omni Adamante,ut terminum proprium habeat, Econtra ut Adamas sit mollior aqua. Ut ex talibus operibus cognoscamus qualis sit noster Deus, nempe Deus omnipotens, qui admirabile coelum ex rudi coelo fecerit, et omnia pro sua voluntate sit operatus* (Eng. trans. Henry Cole, op. cit., p. 49).

others of non-generating and incorruptible principles. Aristotle, in discussing the nature of the heaven, affirms that it is not composed of any elements at all, but has its own peculiar nature".[12]

As we can see from this excerpt, Luther clearly embraces Aristotle's conception of the cosmos and, in this regard, we think that a digression is due.

Luther's commentary was published in 1544 and, as is known, in 1543 the famous work by Copernicus, *De Revolutionibus Orbium Coelestium*,[13] was published. But, some years before, Copernicus had also written a brief sketch (*Commentariolus*[14]) of his astronomical system. The *Commentariolus* was not printed during the life of its author, but a number of handwritten copies circulated for a time among students of science. Moreover, a young professor of the University of Wittenberg, Georg Joachim von Lauchen (called Rheticus after the Roman province of Rhaetia in Austria, where he was born) left his university in the spring of 1539, setting out for Prussia to study with Copernicus. In 1540, he quickly wrote a survey of the principal features of the new astronomy and had it printed at Danzig, in the form of a letter to his former teacher Schöner, with the title *De Libris Revolutionum Narratio Prima G. J. Rhetici ad Joannem Schonerum*.[15] Furthermore, a second edition of the *Narratio Prima* was published in Basel in 1541.

All this is by way of saying that Copernicus's theory was certainly known to Luther before he wrote his Commentary. Written evidence of this has also been found in his famous work *Tischreden* (*Table Talk*). In fact, in a well-known passage of this work, pertaining to 4 June 1539, Luther (without mentioning the name of Copernicus) speaks about a madman who wants to overturn the whole art of astronomy but, as the Holy Scripture shows, Joshua ordered the sun to stop, not the earth.

Philippus Melanchton (1497–1560), a personal friend of Luther and one of the protagonists of the Reformationalso affirmed, in a letter dated 16 October 1541, that Copernicus's attempt at 'moving the earth and stopping the sun' was absurd and that a shrewd government should not tolerate the propagation of such ideas. And, as historians confirm, throughout the sixteenth century the Reformed Church opposed Copernicus's theory. Therefore it is impossible to expect from Luther a discussion of the two verses in the context of the new astronomy.

Let us now take up once more the thread of what we were saying. After having recounted the opinions of the "philosophers", Luther continues: "Moses however proceeds with his narrative of the creation, in all simplicity and plainness; making here *three* divisions of things; waters 'above' the firmament, waters 'under' the firmament, and 'the firmament' in the middle. And in the term heaven, Moses comprehends all that body which philosophers represent by

---

[12] Martin Luther, *In Primum Librum…*, op. cit., p. VIII: *Sed hoc maxime mirabile est, quod Moses manifeste tres partes facit, et firmamentum collocat medium inter aquas. Ego quidem libenter imaginarer. Firmamentum esse supremum corpus omnium, et aquas non supra, sed sub coelis pendentes et volantes, esse nubes quas cernimus, ut sic aquae ab aquis distinctae intelligerentur nubes divisae a nostris aquis in terra. Sed Moses, manifestis verbis aquas supra et infra Firmamenti esse dicit. Quare captivo hic sensum meum, et assentior verbo etiamsi id non assequar. Quaeritur autem hic, Quae sint illae aquae, et qomodo corpora superiora sint distincta: Philosophorum partitio non est ignota. Ponunt enim quatuor elementa, eaque secundum qualitates collocant et distinguunt. Infimum locum terrae, secundum aquae, tertium aeri, postremum et summum igni assignant. Alij his annumerant aethera quintam essentiam. Postea numerantur spherae seu orbes septem Planetarum, et octava sphera stellarum fixarum. Ac convenit fere de his inter omnes, ut sint quatuor spherae generabilium et corruptibilium, Deinde octo aliae generabilium et incorruptibilium. Ac Aristoteles de coeli natura disputat. Quod non sit composita ex elementis, sed suam propriam naturam habeat* (Eng. trans. Henry Cole, op. cit., pp. 49-50).

[13] Nicolai Copernici Torinensis, *De Revolutionibus Orbium Coelestium, libri VI* (Nuremberg: Ioh. Petreium, 1543).

[14] See an English translation of the *Commentariolus* in Edward Rosen, *Three Copernican Treatises* (New York: Dover, 1959).

[15] An English translation of the *Narratio prima* can also be found in E. Rosen, *Three Coperican Treatises*, op. cit.

their *eight spheres*, by *fire* and by *air*. ... And it is manifest that the air in which we live is called, in the Holy Scripture, the heaven; ... This distinction of the spheres therefore is not Mosaic nor Scriptural, but is an invention of men, as an aid to instruction on these astronomical subjects; and which ought not to be despised, as such an assistance.

"But we Christians ought to meditate and think on these things, and their causes, differently from philosophers. And although there are some things which are beyond our comprehension; as for instance these waters that are 'above' the firmament; all such things are rather to be believed, with a confession of our ignorance, than profanely denied, or arrogantly interpreted according to our shallow comprehensions. It behoves us ever to adhere to the phraseology of the Holy Scripture, and to stand by the very words of the Holy Spirit; whom it pleased, in this sacred narrative by his servant Moses, so to arrange the different parts of the great work of creation, as to place, in the midst, 'the firmament'; formed out of the original mass of the unshapen heaven and earth, and stretched out and expanded by the Word: and then to represent some waters as being 'above' that firmament, and other waters 'under' that firmament: both waters being also formed out of the same original rude undigested matter. And the whole of this part of the creation is called, by the Holy Spirit, the heaven; together also with its seven spheres, and the whole region of the air; in which air impressions are made, and in which the fowls wander as they will.

"To return therefore unto the principal matter before us; —When any inquiry is instituted into the nature of these waters, it cannot be denied that Moses here affirms that there are waters 'above' the heavens; but of what kind or nature these waters are, I freely confess, for myself, that I know not: for the Scriptures make no other mention of them than in this place, and in the Song of the three children, in the *Apocrypha*: and I can attempt to declare nothing certain on these and similar subjects. Hence I can say nothing whatever, as known and understood, concerning the heaven where the angels are, and where God dwells with the blessed: nor concerning other kindred things, which shall be revealed unto us in the last day, when we shall have been clothed with another body. —Enough has now been said on this part of the divine subject, to show, that on the *second day* the heaven was so separated, distinguished, and located, that it should be *between* the waters."[16]

The novelty, if such we can call it, introduced by Luther with respect to the former exegetes lies above all in the meaning he proposes to assign to the Latin word *firmamentum* used in the

---

[16] Martin Luther, *In Primum Librum...*, op. cit., p. IX: *Moses de simplici et plano, ut vocant, procedit, et ponit tres partes, aquas supra, et infra, et in medio firmamentum. Ac vocabulo coeli complectitur totum id corpus, quod Philosophi octo spheris, igni et aere, distinguunt. ... Et manifestum est, aerem in quo vivimus, appellari in scriptura sancta coelum, quia scriptura vocat volucres coeli ... Igitur distinctio ista spherarum non est Mosaica, nec sacrae scripturae, sed est ab hominibus eruditis excogitata ad docendum, id quod magni beneficij loco debemus agnoscere... Ergo de causis illarum rerum nos Christiani aliter sentire debemus quam Philosophi. Etsi quaedam sunt supra captum nostrum (sicut ista hic de aquis super coelos) ea potius sunt cum nostrae ignorantiae confessione credenda, quam aut impie neganda, aut arroganter pro nostro captu interpretanda. Oportet enim nos servare phrasin scripturae sanctae, et manere in verbis Spiritus sancti, cui ad hunc modum placuit distribuere Creaturas, ut in medio esset Firmamentum, productum ex isto informi coelo et terra informi, ac per verbum extensum. Deinde supra et infra firmamentum ut essent aquae, etiam ex ista informi massa desumptae. Hoc totum Spiritus sanctus appellat coelum, cum septem spheris, et regione aeris tota, in qua sunt impressiones, et in qua volucres errant ... Ut igitur ad propositam quaestionem redeam. Cum queritur de natura istarum aquarum, negari non potest, quin ut Moses dicit aquae sint super coelos. Cuiusmodi autem aquae illae sint, libere fateor me ignorare. Nam neque scriptura earum meminit usquam praeterit in hoc loco et in Cantico trium puerorum, et nos de similibus rebus omnibus nihil certi possumus praedicare. Sicut nec de coelo in quo Angeli et Deus habitat cum Beatis, nec de alijs quae in novissimo die, cum alia carne induti erimus, revelabuntur, nunc certi aliquid dicere possumus ... Haec de praesenti loco satis sint dicta, quod scilicet secundo die coelum sit distinctum, ut esset medium inter aquas* (Eng. trans. Henry Cole, op. cit., pp. 52-53).

Vulgate. He does not base himself on the direct meaning of the Latin term used in the Vulgate, but goes back to the Hebraic text, drawing from it the meaning of "extension". According to him, the "firmamentum" is not something new and different created on the second day, but the extension of what had already been created in the first day: that kind of initial nebulosity is "inflated" by the Word, as one can do with the bladder of a hog. Luther says that, with this "rustic" similitude (easily understandable by the German audience of that time), the sacred matter may be more plainly understood.

From this, all the rest follows. But the hurdle of the waters "above" the firmament remains. At this point, Luther confesses his incomprehension, or rather his ignorance, and affirms that in any case one must believe in what is written in the Scripture. Hence one has to do with a truth of faith. It must also be noticed that, in his explanations, Luther always appeals to the physical reality which all can see and touch. The only supernatural beings he mentions are the angels, criticizing Moses for failing to mention them in his narrative.

From the time of Augustine the exegetes of the Bible had blamed the problem of the words that are difficult to interprete on a supposed inaccuracy of Jerome's translation. For this reason, as we have seen, Augustine in his maturity studied the Greek and, after him, others referred also to the Greek version. At the time of Luther, the search began again in the original Hebraic version.

3.2 John Calvin (1509–1564)

John Calvin belonged to the generation which followed that of Luther, and he was the second great protagonist of the Protestant Reformation. As we did for Luther, we will dwell neither on his biography nor the meaning of his work, either as a theologian or as an inspirer of political reforms. We shall limit ourselves to the information essential to insert him in the historical context, remaining focussed on the aim we have set in this book.

Calvin was born as Jehan Cauvin in Noyon (France) in 1509. His father had destined him for an ecclesiastic career, but because of a conflict with the ecclesiastic authorities he was excommunicated and was obliged to change his mind. So Calvin first went to the University of Paris, where he graduated master of arts, then to that of Orleans to study law. At the beginning he dealt with law and humanities. It was in the 1530s that he began to adhere to the Protestant Reformation and in 1535, in Basel, he accomplished the first edition of what would be the most significant among his works, *Institutio Christianae religionis* (the definitive edition would be published in 1559).[17]

In contrast to the personalities we have dealt with up to now, Calvin never was a member of a religious order: he was a lay theologian. In the first part of his life he was in various places (Basel, Geneva, Strasbourg, etc.) and, in 1541, he returned to Geneva and remained there until his death in 1564. In this last period in Geneva, Calvin in a certain sense represented the religious conscience of the city. He succeeded in imposing his *Ordonnances ecclésiastiques*, thus giving to the city of Geneva a new constitutional order, even though he never was a member of the Consistory, an ecclesiastical court composed of the lay elders and the ministers.

Let us now speak about the work which Calvin devoted to comment Genesis.[18] Calvin had always had a great interest in the Bible, starting from his study of the French translation of the

---

[17] John Calvin, *Institutio Christianae Religionis, in Libros Quatuor Nunc Primum Digesta ... Aucta Etiam Tam Magna Accessione Ut Propemodum Opus Novum Habere Possit* (Geneva, 1559).

[18] John Calvin, *Commentarii Ioannis Calvini in quinque libros Mosi*. Editio Secunda priori longe emendatior ac locupletior (Geneva: Gaspar. De Hus, 1573). (The first edition dates back to 1563). For the English translation we refer to *The John Calvin Bible Commentaries, Genesis 1-23* (Jazzybee Verlag—Jürgen Beck, 2012).

Bible by Jacques Le Fevre D'Estaples (1455–1536), a presbyter, humanist and theologian who had translated the New Testament (1523) and the Old Testament (1528) and published the complete work at Antwerp in 1530. His translation was based on the Latin text of the Vulgate, and Calvin corrected his translation of the Old Testament making reference to the Hebraic version. So Genesis 1:6-7, the verses which we are interested in, were changed into:

6. *Et dixit Deus, sit extensio in medio aquarum et dividat aquas ab aquis.*

7. *Et fecit Deus expansionem: et divisit aquas quae erant sub expansione, ab aquis quae erant super expansionem. Et fecit ita.*[19]

The work, following the custom of that time, was dedicated to a prominent personality, in this case to "The Most Illustrious Prince, Henry, Duke of Vendome, Heir to the kingdom of Navarre".[20] The peculiarity of the dedication lies in the fact that the dedicatee (the future king Henry IV) was at that time ten years old. Calvin apologizes for not having had the possibility of asking the queen mother's permission to do this but takes it for granted. At the beginning of the dedication he says that the book will not be detrimental for the prince, but he also notes that it may appear incongruous to dedicate a work of such a complexity to a youngster and adds: "Although many things contained in this book are beyond the capacity of your age, yet I am not acting unreasonably in offering it to your perusal, and even to your attentive and diligent study. For since the knowledge of ancient things is pleasant to the young, you will soon arrive at those years in which the history of the creation of the World, as well as that of the most Ancient Church, will engage your thoughts with equal profit and delight".[21]

In any case, almost the entire treatise is used by Calvin to argue against the "Papists", that is, against the Church of Rome, even when he speaks about his work and why he has written it: "I am not ignorant of the abundance of materials here supplied, and of the insufficiency of my language to reach the dignity of the subjects on which I briefly touch; but since each of them, on suitable occasions has been elsewhere more copiously discussed by me, although not with suitable brilliancy and elegance of diction, it is now enough for me briefly to apprise my pious readers how well it would repay their labour, if they would learn prudently to apply to their own use the example of The Ancient Church as it is described by Moses. And, in fact, God has associated us with the holy Patriarchs in the hope of the same inheritance, in order that we, disregarding the distance of time which separates us from them, may, in the mutual agreement of faith and patience, endure the same conflicts. So much the more detestable, there are certain turbulent men, who, incited by I know not what rage of furious zeal, are assiduously endeavoring to rend asunder the Church of our own age, which is already more than sufficiently scattered. I do not speak of avowed enemies, who, by open violence, fall upon the pious to destroy them, and utterly to blot out their memory; but of certain morose professors of the Gospel, who not only perpetually supply new materials for fomenting discords, but by their restlessness disturb the peace which holy and learned men gladly cultivate. We see that with the Papists, although in some things they maintain deadly strife among themselves, they yet combine in wicked confederacy against the Gospel. It is not necessary to say how small is the number of those who hold the sincere doctrine of Christ, when compared with the vast

---

[19] John Calvin, *Commentarii Ioannis Calvini in quinque libros Mosi*, op. cit., p. 1.

[20] John Calvin, *Commentarii Ioannis Calvini in quinque libros Mosi*, op. cit., n.p.: *Illustrissimo Principi Henrico duci Vindocinensi, regni Navarrae haeredi* (Eng. trans. *The John Calvin Bible Commentaries*, op. cit., p. 11).

[21] John Calvin, *Commentarii Ioannis Calvini in quinque libros Mosi*, op. cit.: *Etsi autem aetatis tuae captum superant multa quae in hoc libro continentur: non tamen praepostere tibi legendum offero, et quidem ut in attenta eius lectione sedulo te exerceas. Nam quum adolescentes oblectet rerum veterum cognitio, mox ad eos annos pertinges, in quibus historia tam creationis mundi quam vetustissimae Ecclesiae, non minore cum fructu quam oblectatione te occupet* (Eng. trans. *The John Calvin Bible Commentaries*, op. cit., p. 12).

multitudes of these opponents. In the meantime, audacious scribblers arise, as from our own bosom, who not only obscure the light of sound doctrine with clouds of error, or infatuate the simple and the less experienced with their wicked ravings, but by a profane license of skepticism, allow themselves to uproot the whole of Religion. For, as if, by rank ironies and cavils, they could prove themselves genuine disciples of Socrates, they have no axiom more plausible than, that faith must be free and unfettered, so that it may be possible, by reducing everything to a matter of doubt, to render Scripture flexible (so to speak) as a nose of wax. Therefore, they who being captivated by the allurements of this new school, now indulge in doubtful speculations, obtain at length such proficiency, that they are always learning, yet never come to the knowledge of the truth".[22]

We have dwelt on the dedication (which bears the date 31 July 1563), by quoting two excerpts, in order to emphasize the attitude that Calvin takes towards the Bible. He has not only introduced emendations in the text of the Vulgate, but also uses the Bible as a tool for disputing with the Roman Church.

Let us see how he comments on the works of the second day of the creation and therefore also the question of the waters above the firmament: "The work of the second day is to provide an empty space around the circumference of the earth, that heaven and earth may not be mixed together. For since the proverb, 'to mingle heaven and earth', denotes the extreme of disorder, this distinction ought to be regarded as of great importance. Moreover, the word *yqr* (*rakia*) comprehends not only the whole region of the air, but whatever is open above us: as the word heaven is sometimes understood by the Latins. Thus the arrangement, as well of the heavens as of the lower atmosphere, is called *yqr* (*rakia*) without discrimination between them, but sometimes the word signifies both together sometimes one part only, as will appear more plainly in our progress. I know not why the Greeks have chosen to render the word *stereoma*, which the Latins have imitated in the term, *firmamentum*; for literally it means *expanse*. And to this David alludes when he says that 'the heavens are stretched out by God like a curtain', (Psalm 104:2.). If any one should inquire whether this vacuity did not previously exist, I answer, however true it may be that all parts of the earth were not overflowed by the waters; yet now, for the first time, a separation was ordained, whereas a confused admixture had previously

---

[22] John Calvin, *Commentarii Ioannis Calvini in quinque libros Mosi*, op. cit., p. 3: *Nec me latet quanto uberior hic materia suppetat, et quantum infra eorum quae breviter attigi dignitate subsidat mea oratio: sed quia opportunis locis alibi singula a me, quanquam non quo par erat splendore et ornatu, plenius tamen et copiosus tractata sunt : nunc mihi satis fuit breviter commonefacere pios lectores quantum operae pretium facturi sint, si veteris Ecclesiae exemplar, quale a Mose expressum est, ad suos usus prudenter aptare discant. Et certe ideo nos sanctis Patriarchis in spem eiusdem haereditatis Deus adiunxit, ut superata quae nos separat temporum distantia, mutuo fidei et patientiae consensu eadem certamina obeamus. Quo maiore odio digni sunt turbulenti quidam homines, qui nescio quo furiosi zeli oestro perciti, Ecclesiam nostrae aetatis plus satis dissipatam, assidue discerpere conantur. Non loquor de professis hostibus, qui aperta vi ad perdendum quicquid est piorum, funditusque delendam eorum memoriam incumbunt: sed de morosis quibusdam Evangelij professoribus, qui non modo fovendis dissidiis novam subinde materiam suppeditant: sed pacem, quam pij doctique homines libenter colerent, sua inquietudine perturbant. Videmus ut inter Papistas, quamvis capitalia inter se aliis in rebus certamina exerceant, maneat tamen scelerata contra Evangelium conspiratio. Quam exiguus sit eorum numerus qui sinceram Christi doctrinam tenent, si cum ingentibus conferatur illorum copiis, dicere nihil attinet. Interea emergunt quasi e sinu nostro audaculi scriptores, qui non solum errorum nebulis sanae doctrinae lucem obscurant, vel pravis deliriis simplices dementant, ac minus exercitatos: sed profana dubitandi licentia totam religionem sibi convellere permittunt. Nam quasi putidis suis ironiis et cavillis genuinos Socratis discipulos se probent, nullum illis magis plausibile est axioma quam fidem esse liberam, ut liceat, quavis de re ambigendo, Scripturam instar nasi cerei (ut loquuntur) flexibilem reddere. Itaque hoc proficient tandem qui illecebris novem huius academie capti, nunc dubiis speculationibus indulgent, ut semper discendo, nunquam ad scientiam perveniant veritatis* (Eng. trans. *The John Calvin Bible Commentaries*, op. cit., p. 15).

existed. Moses describes the special use of this expanse, to divide the waters from the waters from which word arises a great difficulty. For it appears opposed to common sense, and quite incredible, that there should be waters above the heaven. Hence some resort to allegory, and philosophize concerning angels; but quite beside the purpose. For, to my mind, this is a certain principle, that nothing is here treated of but the visible form of the world. He who would learn astronomy, and other recondite arts, let him go elsewhere. Here the Spirit of God would teach all men without exception; and therefore what Gregory declares falsely and in vain respecting statues and pictures is truly applicable to the history of creation, namely, that it is the book of the unlearned. The things, therefore, which he relates, serve as the garniture of that theater which he places before our eyes. Whence I conclude, that the waters here meant are such as the rude and unlearned men perceive. The assertion of some, that they embrace by faith what they have read concerning the waters above the heavens, notwithstanding their ignorance respecting them, is not in accordance with the design of Moses. And truly a longer inquiry into a matter open and manifest is superfluous. We see that the clouds suspended in the air, which threaten to fall upon our heads, yet leave us space to breathe. They who deny that this is effected by the wonderful providence of God, are vainly inflated with the folly of their own minds. We know, indeed that the rain is naturally produced; but the deluge sufficiently shows how speedily we might be overwhelmed by the bursting of the clouds, unless the cataracts of heaven were closed by the hand of God. Nor does David rashly recount this among His miracles, that God layeth the beams of his chambers in the waters, (Psalm 104:31) and he elsewhere calls upon the celestial waters to praise God, (Psalm 148:4.) Since, therefore, God has created the clouds, and assigned them a region above us, it ought not to be forgotten that they are restrained by the power of God, lest, gushing forth with sudden violence, they should swallow us up: and especially since no other barrier is opposed to them than the liquid and yielding, air, which would easily give way unless this word prevailed, 'Let there be an expanse between the waters'. Yet Moses has not affixed to the work of this day the note that God saw that it was good: perhaps because there was no advantage from it till the terrestrial waters were gathered into their proper place, which was done on the next day, and therefore it is there twice repeated".[23]

---

[23] John Calvin, *Commentarii Ioannis Calvini in quinque libros Mosi*, op. cit., pp. 4-5: *Opus secundi diei est, inane hoc spatium per terrae circumferentiam, ne caelum terrae misceatur. Nam quum hoc proverbio. Caelum terrae miscere, notetur extrema rakia magni haec distinctio fieri debet. Porro vox* yqr *non modo totam aeris regionem comprehendit, sed quicquid supra nos patet qualiter caelum Latinis interdum capitur. In dispositio caeli et aeris promiscue vocatur* yqr: *sed aliquando utrunque simul, aliquando alteram partem significat: ut melius parebit ex progressu. Nescio cur Graecis placuerit vertere* stereoma: *quod in firmamenti nomen imitati sunt Latini; ad verbum enim est expansio. Atque huc alludit David, quum dicit extensos esse caelos a Deo instar cortinae. Si quis roget annon prius fuerit ista vacuitas: respondeo, utcunque non occuparentur ab aquis omnia: nunc tamen primum ordinatam fuisse distantiam, quum prius incomposita esset permixtio. Usum specialem exprimit Moses, ut distinguat aquas ab aquis, ex quibus verbis oritur magna difficultas. Est enim alienus a sensu communi, et prorsus incredibile, quasdam esse aquas caelo superiore. Quare nonnulli ad allegoriam confugiunt et de Angelis philosophantur: sed intempestive. Nam hoc mihi certum principium est, hic nonnisi de visibili mundi forma tractari: astrologiam et alias artes reconditas aliunde discat qui volet. Hic Spiritus Dei omnes simul sine exceptione docere voluit: atque adeo quod falso et perperam de statuis et picturis Gregorius pronuntiat, vere in hanc creationis historiam competit, librum scilicet esse idiotarum. Ergo quaecumque commemorat, ad ornamentum illius theatri spectant quod nobis ante oculos ponit. Unde statuo, aquas hic intelligi quas rudes quoque et indocti perspiciant. Quod enim quidam se fide amplecti dicunt quod hic legunt de aquis supercaelestibus, utcumque eas ignorent, Mosis instituto non est consentaneum. Et vero rei apertae et expositae supervacua est longior inquisitio. Videmus nubes in aere suspensas capitibus nostris sic minari, ut spirandi locum nobis relinquant. Hoc mirabili Dei providentia fieri qui negant, frustra ingenii sui vanitate sunt inflati. Scimus quidem naturaliter procreari pluvias: sed diluvium satis ostendit quam mox obruendi simus nubium impetu, nisi cataractae caeli, Dei manu clausae essent. Nec temere inter eius miracula hoc recenset David, quod in aquis ponat Deus coenaculorum suorum trabes. Psal. 104. a. 3. et alibi caelestes aquae ad laudandum Deum advocat. Quum*

We can say that Calvin's commentary on the works of the second day follows Luther's example in adopting the paradigm of the "expanse" for the firmament (the same paradigm, as we shall see, will be also adopted by Pererius), but represents an innovation as regards the "materiality" of the waters. Calvin is very firm in maintaining that what was written by Moses is a description of the world as it can be observed by everyone; in fact he adds that the Bible is the book of the unlearned (*librum scilicet esse idiotarum*). That is, for Calvin the water is true water and not a metaphor and, to justify that this water is sustained by the "expanse", he asserts that the clouds that all can see are also suspended above us by the hand of God. Therefore even if, as he acknowledges, "it appears opposed to common sense, and quite incredible", the superior waters are sustained by the "expanse", but through the hand of God. Therefore, at the end, the recourse to the power of God is reiterated.

To have, so to speak, made Moses's description banal by bringing it back to a description of the everyday world has not completely solved the problem: the hand of God always turns out to be indispensable.

3.3 Benedictus Pererius (1535 - 1610)

We now go on to a Catholic annotator: Benedictus Pererius. We have chosen to call the author of the commentary we shall deal with by using the Latin name which appears in the title of his works in order to avoid misunderstandings. At present he is called Benet Perera,[24] but in the past he was also called Pereira. As far as it is known, he was born at Ruzafa, near Valencia (Spain) in 1535 and entered the Society of Jesus in 1552. Biographical notes regarding his life before this date are not available. Afterwards he was sent (by the Order) first to Sicily and then to Rome to complete his education. In Rome, he joined the Collegio Romano where over the years, starting in 1558, he taught various courses in several disciplines (physics, metaphysics, logic,etc.).

During his activity at the Collegio Romano, he took part to several philosophical disputes which sometimes degenerated into heated arguments. It is known that he was engaged in a debate with Christopher Clavius on the nature of mathematics. Also, his relations with the Averroist philosophers of the University of Padua were criticized by the authorities of the Church.

The most important among his works is considered to be *De communibus omnium rerum naturalium principiis et affectionibus libri quindecim*,[25] written to compete with the Renaissance natural philosophers who were his contemporaries. Another work, particularly important for our research, is the commentary he wrote on Genesis, which appeared in the first

---

ergo nubes creaverit Deus, et regionem illis supra nos assegnaverit, non debuit praeteriri, Dei virtute arceri ne profusae subito impetu nos absorbeant ac praesertim quum non aliud illis oppositum sit claustrum quam liquidus et evanidus aer, qui facile cederet nisi praevalerat hoc verbum, Sit extensio inter aquas. Caeterum huius diei operi notam illam non opposuit Moses, quod Deus esse bonum videri: forte quia nondum extabat utilitas, donec aquae terrestres in proprium locum se reciperent: quod postero die factum est: itaque bis illic repetitur (Eng. trans. *The John Calvin Bible Commentaries*, op. cit., pp. 31-32).

[24]We refer to the proceedings of the conference *Benet Perera (Pererius, 1535–1610). Un gesuita rinascimentale al crocevia della modernità / A Renaissance Jesuit on the Threshold of Modernity* (Rome, 13-14 dicembre 2013, Pontificia Università Gregoriana, Scuola Normale Superiore, Università di Chieti e Pescara) (Brepols, 2015). Particularly interesting, in the context of our research, is the communication of Paul Richard Blum, "Platonic References in Pererius' Comments on the Bible".

[25]Benedicti Pererii Societatis Jesu, *De communibus omnium rerum naturalium principiis et affectionibus libri quindecim* (Paris: Michel Sonnius, 1579).

edition in 1589 with the title *Prior Tomus Commentariorum et Disputationum in Genesim*[26] and was successively augmented and completed in the ensuing editions. This is a work which, even in its first edition, is subsequent to the rules issued by the Tridentine Council with regard to the commentaries on the Bible. Therefore it assumes a particular importance for understanding how an annotator—in the case in point a Jesuit of the Collegio Romano—had to move (or, better, could move) by avoiding the obstacles placed by the Council.

The work was dedicated to the cardinal Enrico Caetani. After the Preface, Pererius reports the text of the two chapters of Genesis in the version of the Vulgate and enunciates the four rules which (according to him) must be followed in the commentaries on Genesis: "Four rules are presented with which it is easy to judge which among the interpretations the various authors have given of Moses's doctrine of the generation of the world is true or false and which interpretation is more or less probable".[27] The first rule dismisses the allegorical interpretation. In fact it says: "Moses's doctrine, which recounts the creation of the world, is certainly historical" (*est plane historica*) and relies on the work of Augustine and of other Fathers of the Church. In the second rule he censures the recourse to miracles and the omnipotence of God to explain the things when it is not necessary. Some, he says, seek refuge in this expedient as in an asylum (*quasi ad asylum*). In the third rule he warns the annotators against becoming enamored of their idea and defending it even with mordacity and intransigence (*non modo teneat mordicus et praefractae*). Finally, the fourth rule is the one which Galileo quotes in the famous letter to the Grand Duchess Christina, together with the sentence of St. Augustine we have already mentioned in Section 1.6. As is known,[28] Galileo, in his early years, studied the textbooks of the courses of physics run at the Collegio Romano, but there is no evidence that he had a copy of Pererius's Commentary on Genesis in his personal library.[29] The fourth rule says: "We must also take heed, in handling the doctrine of Moses, that we altogether avoid saying positively and confidently anything which contradicts manifest experiences and the reasoning of philosophy or the other sciences. For since every truth is in agreement with all other truth, the truth of Holy Writ cannot be contrary to the solid reasons and experiences of human knowledge".[30]

And now, finally, let us see how Pererius deals with the works of the second day, applying his rules. He begins by presenting the difficulties inherent in their interpretation: "This passage presents two questions which are difficult and, in the opinion of some, inexplicable; in our opinion [they are] not yet quite sufficiently explained. The first question is what Moses would signify with the word *firmamentum* and the second is what we must understand for the waters which, according to Moses, are placed above the firmament. We shall briefly, though diligently

---

[26] Benedicti Pererii Valentini e Societate Iesu, *Prior Tomus Commentariorum et Disputationum in Genesim* (Rome: Georgium Ferrarium, 1589).

[27] Benedicti Pererii Valentini, *Prior Tomus*, op. cit., p. 10: *Traduntur Quatuor Regulae, quibus facili indicari possit, de variis auctorum hanc Mosis de generatione Mundi doctrinam explanantium interpretationibus, qua vera aut falsa sit, et qua sit magis minusque probabilis interpretatio* (our Eng. trans.).

[28] On this there are studies by A. Carugo, A. C. Crombie, W. A. Wallace.

[29] See Antonio Favaro, "La Libreria di Galileo Galilei descritta e illustrata da Antonio Favaro", Bullettino di Bibliografia delle Scienze Matematiche, vol. XIX (1886).

[30] Benedicti Pererii Valentini, *Prior Tomus*, op. cit., p. 12: *Illud etiam diligente cavendum est et omnino fugiendum est, ne in tractanda Mosis doctrina quidquam affirmate et asseveranter sentiamus et dicamus, quod repugnet manifestis experimentis et rationibus philosophiae vel aliarum disciplinarum: namque, cum verum omne semper cum vero congruat, non potest veritas Sacrarum Literarum veris rationibus et experimentis humanarum doctrinarum esse contraria* (Eng. trans. by Stillman Drake, from *Discoveries and Opinions of Galileo* (Garden City, NJ: Doubleday, 1957), p. 186).

and accurately, expose our opinion, and the opinion of others, on these questions".[31] Actually he does not respect what he promised with the adverb "briefly" since he takes fourteen pages to expose his opinion. He begins by showing how the word *firmamentum* had been interpreted in the past (interpretations which we already know) and then starts: "What is then our opinion on the firmament? We hold that the true and genuine meaning of *firmamentum*, the sense of these words of Moses, must be pulled out from the Hebraic word in this passage. That which the Latin interpreter (St. Jerome) translates *firmamentum,* and the Septuagint in Greek *stereoma,* in Hebraic is *Raquiah,* a word which properly means extension, from the verb *Racah,* which means to expand, or to extend: in fact with this meaning it is frequently used in the Sacred Scriptures, as one can verify in Isaiah 42, Psalms 133, and Jeremiah 10, and in other writings, where the Hebraic word means to extend, though the Latin interpreter, following the Septuagint, have often expressed it with the word *firmare*".[32]

At this point, Pererius lists all the ways in which an extension can be realized and continues to quote passages of the Sacred Scripture as a confirmation. He then says: "We therefore have the opinion that Moses in this passage with the name *firmamentum* means all that space which expands around the earth until the stars: obviously as far as a keen eyesight can go. The extreme part of this space is held by the heavens and the stars, the deepest part by the element of the fire and the air".[33]

As Arnold Williams notices, Pererius, "though he cannot flatly object to the authoritative Vulgate translation, likewise interprets the word in the sense of expanse".[34] The expedient, as we have seen, has been executed by tracing in the Sacred Scripture various passages which somehow refer to an expansion.

Then he passes to the second question: "Having exhausted the first question of the firmament of the heaven, we prepare ourselves to deal with the other, that of the waters placed above the firmament, even more obscure and controversial. What those waters were it seemed extremely difficult to explain to those who held that in that passage with the name of firmament one would indicate the sidereal heaven: for this reason, they accounted for many interpretations of those waters, and others for other interpretations, not only different, but even opposite".[35]

---

[31] Benedicti Pererii Valentini, *Prior Tomus,* op. cit., p. 57: *Duas hic locus quaestiones habet perquam difficiles, et quorumdam iudicio inexplicabiles; ut autem mea fert sententia, nequaquam adhuc satis explicatas. Prior quaestio est, quidnam Moses nomine firmamenti significare voluerit: posterior autem quaestio est, quales aquas intelligere oporteat, quas Moses ait supra firmamentum esse locatas. De his quaestionibus, et aliorum iudicium, et nostram sententiam, breviter quidem, diligenter tamen et accurate exponemus* (our Eng. trans.).

[32] Benedicti Pererii Valentini, *Prior Tomus,* op. cit., p. 60: *Quae igitur est nostra de firmamento sententia? Arbitramur veram firmamenti significationem et germanam propriamque; horum verborum Mosis sententiam ex voce hebraea, qua est hoc loco, esse eruendam. Quod latinus Interpres reddit, firmamentum, et LXX interpretes graece verterunt* stereoma, *hebraice est* Raquiah, *quae vox proprie significat extensionem, a verbo* Racah, *quod est expandere, seu extendere: haec enim significatione saepenumero usurpatur in Sacris Litteris, ut animadvertere licet Isa.42. Psal.133. et 136. et Hierem. 10., alijsque. locis, in quibus est verbum hebraicum significans extendere, quod tamen latinus interpres imitatus LXX, frequenter expressit verbo firmare* (our Eng. trans.).

[33] Benedicti Pererii Valentini, *Prior Tomus,* op. cit., p. 60: *Nos igitur existimamus nomine firmamenti, hoc loco significari a Mose totum illud spatium, quod expansum et diffusum est circum terram usque ad astra: scilicet quatenus oculorum acies usquequaque porrigi potest. huius spatii supremam partem tenent caeli et astra; imam partem elementum ignis et aeris* (our Eng. trans.).

[34] A. Williams, *The Common Expositor,* op. cit., p. 55.

[35] Benedicti Pererii Valentini, *Prior Tomus,* op. cit., p. 63: *Expedita priori quaestione de firmamento caeli, alteram magis etiam perplexam et litigiosam de aquis supra firmamentum positis, tractare aggredimur. Quales essent illae aquae, difficillimum explicatu visum est ijs quibus nomine firmamenti caelum sydereum hoc loco significari erat persuasum: quamobrem multas illi earum aquarum interpretationes, et alias alij non diversas modo, sed adversas etiam tradiderunt* (our Eng. trans.).

Pererius, following the proven scheme inaugurated by Aristotle in the *Metaphysics*, reviews all the different explanations given through the centuries by the annotators who preceded him, as Aristotle did with the theories of the pre-Socratics. He starts from Origen (*Incipiam ab Origen*) and arrives at his close predecessors. Then, finally expounds his opinion: "There remains little to be cleared and then to explain our opinion (in fact, having refuted others opinions, only this remains to be done). To be sure, we have not been the first to find it, but even so we have approved it against the other opinions: so the explanation and demonstration of it will be easier and direct. And, when we progressed a little about the firmament, we have shown clearly enough that Moses, in that passage, with the name of firmament did not want to indicate the sidereal heaven, but the airy space which expands around the earth and spreads out until the stars. Therefore we think that the waters which are above the firmament are nothing other than those which arise on the air, in that region which philosophers call the middle region of the air. In fact, if in this passage the firmament is the airy space which interposes between the earth and the stars, beyond question it is right to interpret the waters which are above the firmament as those which arise in the higher part of the air: in fact there the vapors released from the earth and from the waters by the power of the sun condense and gather in clouds, and the rains they bring forth run so much profusely and opportunely for irrigating and fertilizing the land, sometimes so abundantly and swiftly that they generate lakes and rivers, if not seas. And for this, as the Scripture tells, the cataracts of heaven were opened in order that the deluge could take place, and by this one means very heavy rains plunging downwards with a greatest force and roar. What Aristotle tells in his first book of *Meteorology*, III, Chap.1 (that, between the heaven and the earth, an immense river mixed of air and water is perpetually carried alternatively upwards or downwards) can help us to understand this; in fact near the rising sun the river of vapor flows upwards whereas that of water moves away downwards. And this must occur in perpetual order".[36]

After this, Pererius continues, digressing at length, to review some of the solutions of the question of the waters above the firmament given through the centuries, starting with a quotation from the Proverbs of Solomon. Finally, he concludes: "Certainly the time has arrived

---

[36] Benedicti Pererii Valentini, *Prior Tomus*, op. cit., pp. 68-69: *Restat ut paucis aperiamus, et expediamus (confutatis namque; aliorum opinionibus hoc unum est reliquum) sententiam nostram; non quidem a nobis primo inventam, sed nobis tamen praeter caeteras probatam: cuius eo facilior atque; promptior erit explicatio et probatio, quod paulo superius cum egimus de firmamento, satis perspicue ostendimus, nomine firmamenti hoc loco Mosen non caelum sydereum, sed spatium aereum quod circum terram usquequaque diffusum est et usque ad sydera expansum, significare voluisse. Nos igitur existimamus aquas quae sunt super firmamentum, non esse alias quam quae in sublimi aere, quam mediam regionem aeris vocant philosophi, generantur. Si enim firmamentum hoc locum, est spatium aereum quod interpatet inter terram et sydera, proculdubio aquas quae supra firmamentum sunt, non alias interpretari convenit, quam quae in superiori parte aeris gignuntur: illuc enim vapores ex terra et aquis, potentia solis elati, densantur et coguntur in nubes, ex quibus imbres generati, tam large et commode ad irrigandam et foecundandam terram defluunt, tanta nonnumquam copia et impetu, ut videantur illic esse ingentes lacus et flumina, quin etiam maria. Quocirca ut fieret diluvium, Scriptura narrat referatas esse caeli cataractas, quo significantur densissimi imbres maximo impetu et fragore deorsum praecipitati. Ad hoc intelligendum pertinet, quod tradit Aristoteles I. libro Meteororum summa 3. cap.1. inter caelum et terram ferri per aera ingens quoddam flumen sursum ascendens, et deorsum descendens perpetua vicissitudine. Verba eius sic habent, Oportet autem intelligere, hunc veluti fluvium, fluere circulariter sursum et deorsum, communem aeris et aquae, prope enim existente sole, vaporis sursum fluit fluvius : cum autem elongatur, aquae deorsum. Et hoc perenne vult fieri, secundum ordinem* (our Eng. trans.) The actual collocation of the passage from Aristotle is *Meteorology* I, 9.

for us to finish here this debate on the supracelestial waters which, precisely because of its great complexity and the variety of opinions, we were obliged to lengthen beyond our will".[37]

In conclusion, we can say that the Jesuit Pererius pursues the work of the Protestant Calvin. The key to all that lies in the returning to the Hebraic version of Genesis and in translating, differently from St. Jerome, the word he had translated *firmamentum* (with the meaning deriving from the Latin verb *firmare*, that is, to consolidate). Once it has been established that this term depicts what has been formed with the expansion of the *caelum* created on the first day and identifying it with the airy space between the earth and the stars, all falls into "normality". Having already accepted the idea that Moses wrote Genesis in a form understandable by the unlearned people, the new translation, which puts everything once again in the order that the man in the street can also ascertain personally, allows one to have an interpretation which we might define of "good sense". And Pererius achieved his goal despite the rule imposed by the Tridentine Council.

3.4 David Pareus (1548–1622)

Until now, we have been interested in Luther, Calvin and the Jesuit Pererius. Now we shall deal with the work of a Lutheran who became a Calvinist and was always engaged in disputes with the Jesuits: David Pareus.

Pareus was born at Frankenstein (in Schlesien) in 1548. When still very young he was apprenticed to an apothecary and then to a shoemaker. At sixteen he entered the school of Christopher Schilling at Hirschberg and, under the influence of his master, abandoned Lutheranism and became a Calvinist (and his father disinherited him for this). Afterwards, in 1566, Schilling made him enter the Collegium Sapientiae at Heidelberg, where he studied theology under the guidance of Zacharias Ursinus. He was appointed professor in this Collegium in 1584 and became its director in 1591. In 1598 he entered the theological faculty as a teacher of the Old Testament and from 1602 until his death he taught the New Testament. As a teacher he had a good reputation but often ran into trouble (causing him to move) due to the political events of that time. He finally came back to Heidelberg, where he died on June 1622. From the point of view of his literary activity, Pareus was very prolific and held several theological debates, particularly against the Jesuits, whom he treated as heretical.

We are obviously interested in his work of commentary on Genesis: *In Genesin Mosis Commentarius*.[38] In the subtitle of the book, inter alia, Pareus asserts that the interpretations of the Jesuits Robert Bellarmino and Benedictus Pererius are refuted (*solide refutantur*). But this, as we shall see, does not concern the waters above the firmament. In any case, he had an axe to grind against the Jesuits. In fact he even wrote several books to harshly criticize a work by Bellarmino.[39]

Let us now deal with the *Commentarius*. It is a work which clearly aspires to establish an interpretation (from the point of view of the Calvinist religion) of Genesis that is as complete and exhaustive as possible, with evident teaching purposes. The book is voluminous, in *quarto*, and contains more than a thousand pages (2254 columns, without considering the two copious tables of contents and the *Epistola Dedicatoria*). One of the tables is called *Elenchus Dubiorum*

---

[37] Benedicti Pererii Valentini, *Prior Tomus*, op. cit., p. 71: *Verum iam tempus est ut hanc de supercaelestibus aquis disputationem, quam propter eius magnam obscuritatem sententiarumque varietatem, longius quam voluissemus producere necesse fuit, tandem hoc loco terminemus* (our Eng. trans.).

[38] David Pareus, *In Genesin Mosis Commentarius* (Frankfurt, 1609).

[39] David Pareus, *Roberti Bellarmini,...Liber Unus De Gratia Primi Hominis-Explicatus et Castigatus Studio Davidis Parei* (Heidelberg, 1612).

et Quaestionum explicatarum hoc Commentario. At the item *De Operibus II Diei – Expanso et divisione aquarum,* we find listed as *Dubia* the two problems we are interested in:

31 What one means by the name *expanse* or *firmament*: the only eighth sphere, or regions of the air: or celestial spheres: or the whole structure from the earth till the extremity of the vault of the universe?

32 In what way the waters under and above the expanse have been divided: and it is shown that the waters above the expanse are clouds.[40]

Let us start with an excerpt regarding the first question: "What was then this expanse? All that, which now is, or one can perceive or can anyway be experienced, from the surface of the earth and of the waters till the extremity of the vault of the universe. Before the second day, outside the surface of the waters there was nothing either empty or full, as not even before the first day there was something both empty and full here, where in the first day the earth and the abyss were created. Now, really we see, outside and around the surface of the earth, that it exists an immense and round space and in this space spherical celestial bodies which move indefinitely according to a certain law; and under the celestial spheres the air around the earth and all things down revolve. Therefore all this is undoubtedly denoted with the name of expanse. FIRST, The space itself, or better the circular motion of the universe, which now exists, but which there was not before the first day outside or around the surface of the abyss, not because there was no space or emptiness before the things, but because the things could not exist without the space. Therefore the space is created together with the things, as the place with the things placed…".[41]

After this, Pareus goes into a pedantic presentation in which Ptolemy's doctrine and the sphericity of the celestial bodies are also quoted, but having learned what, in his opinion, the nature of the expanse is, we pass directly to the second question. Let us begin with a second excerpt: "*Et divisit Deus aquas quae etc.* An all-wise architect, God does nothing without reason. Therefore he also made the expanse of the heavens for an evident use, in order to make it divide the waters from the waters, the superior ones from the inferior ones. In fact he divided the waters by means of the expanse made in the middle of the waters. What was indeed the purpose of the division of the waters? FIRST: In order to diminish the depth of the abyss, in which the earth floated submerged and make it emerge little by little. THEN: in order to clearly manifest his wisdom and power with the work which held up the waters, heavy by nature, above the expanse without support. But one asks what this division was and of what waters and what those waters above the expanse are. I reply: First God speaks openly of the division of the waters of the abyss, when he says: *Sit expansum in media aquarum etc.* In fact no other water

---

[40] David Pareus, *In Genesin Mosis Commentarius*, op. cit., n.p. (Index) : *31 Quid nomine expansi seu firmamenti intelligatur: an sola octava sphaera, an regiones aeris: an orbes coelestes: an tota compages a terra sursum ad extremum universi convexum?*
32 Quomodo divisae fuerint aquae infra et supra expansum: et quod aquae supra expansum sint nubes ostenditur (our Eng. trans.).

[41] David Pareus, *In Genesin Mosis Commentarius*, op. cit., col. 163: *Quid igitur fuit hoc expansum? Totum illud, quod nunc est, vel cernitur aut quovis modo sentitur esse inde a terrae et aquarum superficie usque ad extremum universi convexum. Ante diem secundum extra superficiem aquarum nihil fuit neque vacuum neque plenum: sicuti nec ante primum diem quicquam fuit vel vacuum vel plenum ibi, ubi primo die creata est terra et abyssus. Nunc vero extra et circa superficiem aquarum et terrae videmus esse immensum et rotundum spacium, in hoc spacio corpora coelestia sphaerica, indefiniter certa lege mobilia: et infra sphaeras coelestes aerem terrae circumfusum et omnia haec inferiora proxime ambientem. Hoc igitur totum haud dubie nomine expansi huius significatur: nimirum PRIMO spacium ipsum seu ambitus universi, qui nunc est, sed ante diem primum non fuit extra vel circa superficiem abyssi: non quod fuerit spacium aut vacuum ante res: sed quia res non potuerunt esse sine spacio; proinde spacium necessario intelligitur rebus concreatum, sicut locato locus* (our Eng. trans.).

was created at that time. Hence he divided these waters, but in which way? Absolutely not through the center, as in the cutting of an apple in two equal parts, but in the middle, or better in the midst of the waters by making an expanse of a spherical shape as if one, say, in a ball or in an onion of twelve shells, expanded the six external shells of the spherical space by separating them from the six internal. But it is difficult to find something similar to this skill of God. Hence God divided the waters under the expanse from those above the expanse: that's all".[42]

Pareus continues by describing the division operation as he imagines it, and then comments: "This was the wonderful work of the divine wise and omnipotence: through it, with part of the waters above the expanse and part under it, it was made a large part of the waters move from the earth. From this one can not only grasp what and how useful that division of waters has been, but also what those waters above the expanse are, about which so much by many has been scrupulously discussed".[43]

He then maintains: "*Aquas suspensas esse nubes*" (The highest waters are clouds) and takes seven points to disprove the other interpretations. The seventh point begins with the following words: "And finally all other interpretations are ignorant and absurd, witness Luther: and it is not necessary to refute them one by one".[44]

Instead, he continues to argue about them and, at the end, he concludes: "So then we resume: anything which is above part of the expanse is correctly said to be above the expanse, since the expanse is for Moses the part and the whole, as it is manifest. So the clouds are above the lower part of the expanse. In fact they soar in the middle region of the air above the lower one. Therefore it is correct to say that the clouds are above the expanse".[45]

We can say that Pareus's interpretation does not differ, in substance, from that by Pererius. The waters above the expanse are the clouds and the expanse is air and not the starry heaven. There is no longer any suggestion of an allegorical interpretation. It is a completely "lay" interpretation.

---

[42] David Pareus, *In Genesin Mosis Commentarius*, op. cit., cols. 167-168: *Et divisit Deus aquas quae etc.* Sapientissimus architectus Deus nihil facit frustra. Expansum igitur coelorum etiam ad usum manifestum fecit, ut divideret aquas ab aquis, superas ab inferis. Divisit enim aquas per expansum factum in medio aquarum. Ad quid vero aquarum divisio? *Primum* ad minuendam abyssi profunditatem, in qua terra demersa fluitabat, ut paulatim emergeret. *Deinde* ad declarandam sapientiam et potentiam opificis, aquas natura graves sustentantis supra expansum, absque fulcro. Sed quaeritur, quarum aquarum et qualis fuerit ea divisio, et quae sint illae aquae super expansum? *Respondeo:* Primum de divisione abyssi Deum loqui apertum est, cum dicit: *Sit expansum in medio aquarum etc.* Has igitur aquas per expansum divisit, quomodo? Non utique per centrum, ut cum secatur pomum in duo aequalia: sed in medio, seu intra medias aquas faciendo expanso orbiculariter: ut si quis verbi causa in pila vel caepa duodecim tunicarum, sex tunicas exteriores undiquaque distentas spacio orbiculari intrinseco a sex interioribus separando expanderet. Sed vix invenio simile satis adumbrans hoc Dei artificium. Divisit igitur aquas sub expanso ab aquis super expansum: hoc est (our Eng. trans.).

[43] David Pareus, *In Genesin Mosis Commentarius*, op. cit., col. 168: *Hoc mirandum fuit divinae sapientiae atque omnipotentiae opus: quo factum est, ut aquae partim supra, partim infra expansum ab invicem divisae, terraque magna iam aquarum parte levata esset. Unde intelligitur non modo qualis quamque utilis fuerit ea aquarum divisio: sed etiam. quae sint illae aquae supra expansum : de quibus adeo scrupulose a multis est disputatum* (our Eng. trans.).

[44] David Pareus, *In Genesin Mosis Commentarius*, op. cit., col. 170: *Denique omnes interpretationes aliae sunt ineruditae, et absurdae, Luthero teste: nec est opus refutatione singularum* (our Eng. trans.).

[45] David Pareus, *In Genesin Mosis Commentarius*, op. cit., col. 172: *Sic igitur colligimus: Quicquid est supra partem expansi: id supra expansum recte dicitur: quia expansum est Mosi pars et totum, ut patet. Atqui nubes sunt supra partem infimam expansi. Haerent enim in media regione aeris supra infimam. Ergo nubes supra expansum recte dicuntur* (our Eng. trans.).

3.5 Cornelius à Lapide (1567 - 1637)

Let us pass now to deal with Cornelius à Lapide, an author whom we could a priori consider not unlike Pererius: like him a Jesuit, like him a teacher at the Collegio Romano in the final part of his life. However, as we shall see, at least from the point of view of the interpretation of the three verses, he represents without any doubt a partial return to the old interpretations, in spite of the fact that he wrote a generation after Pererius.

Let us supply now some short biographical notes on him, beginning with his peculiar family name. Of Flemish origin, he was christened Cornelius Cornelissen van den Steen. Since his works were written in Latin, the name *van den Steen* (from stone) was literally translated into Latin: *à Lapide*. He was born at Bocholt, near Liege (Belgium). He studied humanities and philosophy at the Jesuit colleges of Maastrict and Cologne and then theology first at the University of Douai and afterwards at the University of Leuven. He entered the Society of Jesus on 11 June 1592 and was ordained a Catholic priest on 24 December 1595. Afterwards he became professor of Sacred Scripture at Leuven in 1596, and the next year also professor of Hebrew. On 3 November 1616 he was called to Rome to teach the same disciplines at the Collegio Romano, where he remained until his death (on 12 March 1637).

The principal work of Lapide was a project of commentaries on all the books of the Catholic Canon of Scripture, which remained unfinished at his death, even though it amounted to several volumes. One of the first books he did publish was *Commentaria in Pentateuchum Mosis*,[46] first published at Antwerp in 1616 and many times reprinted.

From this work we shall take the comment on the *De Opere secundae diei*. He begins by recalling that the *firmament*, in the Hebrew version, was called with a word deriving from a verb which means to expand or to extend but he immediately adds: "...and, by extending, to fix and to consolidate a thing which before was fluid and loose. As molten copper expands and condenses, in the same way here the water condensed in the heavens, in Greek is named *stereoma* and in Latin *firmamentum*: in fact the firmament is like a wall in the midst of the waters, that is, between the two waters, just interposed between the superior ones and the inferior ones, by separating and compressing them alternatively. One will ask: 'what is this firmament?'".[47] At this point, starting from Origen and ending with Pererius and Gregory of Valencia (he too a Jesuit of the Collegio Romano), he briefly reports the most popular interpretations in the past and then speaks his mind: "As a matter of fact, I say that the firmament is the starry heaven and all those nearby celestial bodies, both inferior and superior until the empyrean".[48]

And, regarding the water: "...it is transformed and consolidated into heavens and air as I said: the thinner and finer part of it remained above heavens, the coarser under the firmament is divided into seas and rivers. And thus, above all heavens, and until above the *primum mobile* (the firmament encompasses all these things) immediately under the empyrean heaven, there

---

[46] Cornelius à Lapide, *Commentaria in Pentateuchum Mosis* (Antwerp: Nutius & Meursius, 1616).

[47] Cornelius à Lapide, *Commentaria in Pentateuchum Mosis, ultima editio aucta et recognita* (Antwerp, 1681), p. 40: *...et distendendo firmare ac solidale rem aliquam, quae prius fluida erat et rara. Sicuti ergo aes fluidum fundendo distenditur et condensatur: ita hic aqua densata in caelos, vocatur graece* stereoma, *latine firmamentum: est enim firmamentum quasi murus in medio aquarum, id est inter duas aquas, superas scilicet et inferas interjectus, easque ab invicem dispescens et coercens. Quaeres, quodnam sit hoc firmamentum, et quaenam sint aquae supra firmamentum?* (our Eng. trans.).

[48] Cornelius à Lapide, *Commentaria in Pentateuchum Mosis* (1681), op. cit., p. 40: *Verum dico, firmamentum esse caelum stellarum, omnesque orbes caelestes illi vicinos, tam inferiores, quam superiores usque ad empyreum* (our Eng. trans.).

are true and natural waters".[49] He provides this reason: "Because this emerges from Moses's simplest and historical narrative. In fact *firmamentum*, and the Hebrew *rakia*, do not mean air or clouds, but properly starry heaven and celestial bodies".[50] Below he finally says that God has placed those waters above the heavens in order to adorn the universe and possibly also for the pleasure of the saints who exist in the empyrean heaven. Here emerges a scrap of the old allegorical interpretations, where the saints were present! At the end, as in the old interpretations (Isidore), he reprises the argument on the etymology of the word *caelum* from the verb *caelare*. *Nihil sub sole novum*!

Lapide, as the saying goes, seems to have his cake and eat it. That is, he complies with the rule established by the Tridentine Council, but at the same time shows that he is aware of the Hebrew version (*rakia*) which suggests "expanse" instead of "firmament".

3.6 Marin Mersenne (1588–1648)

We shall deal now with an author deeply different from those we have been interested in up to now. As Arnold Williams has remarked, "Indeed, Mersenne's name shines much brighter as a scientist than as an exegete".[51] Therefore, the right thing to do would be to dedicate to Mersenne a scientific biography and not one merely consisting of dates and places. But, as in all other cases, we shall limit ourselves to this last kind of biography.[52]

Marin Mersenne was born in La Soultière, near Oizé (France), in 1588 and was educated at Le Mans and then at the Jesuit college of La Flèche. After two years of theology at the Sorbonne, he entered the religious order of Minims with the task of teaching philosophy. He taught both in Nevers and in Paris, settling in the convent of L'Annonciade, where he remained, except for some brief trips abroad, until his death in 1648.

As we have alluded above, Mersenne was above all a scientist who dealt with different branches of science (mechanics, acoustics, optics, geometry, number theory) as well as musical theoretics. In mathematics a particular category of prime numbers is known as "Mersenne numbers". But, beyond the scientific results he obtained, Mersenne played an important role in the scientific milieu of the first half of the seventeenth century as a cultural organizer. In fact he came into contact with most of the prominent scientists of that time and entertained with them an uninterrupted correspondence.[53] Among his correspondents we can cite Descartes (with whom he was a personal friend), Huygens and Galileo (of whom he even translated two works[54]). His activity, which fostered encounters of the greatest scientists of the day, embodied

---

[49] Cornelius à Lapide, *Commentaria in Pentateuchum Mosis* (1681), p. 41: ...*conversa est et consolidata in caelos et aerem ut dixi: pars subtilior et nobilior supra caelos remansit: pars crassior subtus firmamentum in maria et flumina varia divisa est. Itaque supra caelos omnes, adeoque supra primum mobile (haec enim omnia complectitur firmamentum) proxime sub caelo empyreo, sunt verae et naturales aquae* (our Eng. trans.).

[50] Cornelius à Lapide, *Commentaria in Pentateuchum Mosis* (1681), p. 41: *Quia hoc exigit simplicissima et historica Mosis narratio. Firmamentum enim, et Hebraeum rakia, non aerem, non nubes, sed proprie caelum sidereum, orbesque caelestes significat* (our Eng. trans.).

[51] A. Williams, *The Common Expositor*, op. cit., p. 179.

[52] One can find an accurate and up-to-date scientific biography in the *Stanford Encyclopedia of Philosophy* (https://plato.stanford.edu/entries/mersenne/), with a rich bibliography.

[53] This enormous quantity of letters has been published in *Correspondance du P. Marin Mersenne*, Cornelius De Waard, René Pintard, and Bernard Rochot, eds., 17 volumes (Paris: G. Beauchesne, 1932-1988).

[54] See *Les Mechaniques de Galilé, Mathématicien et Ingenieur du Duc de Florence....Traduites de l'Italien par L. P. M. M.*, (Paris, 1634) (critical ed. Bernard Rochot, Paris, P.U.F., 1966) and *Les nouvelles Pensée de Galilei, Mathématicien et Ingenieur du Duc de Florence...Traduit d'Italien en Francois* (Paris, 1639) (critical ed. Pierre Costabel and Michel-Pierre Lerner, Paris, Vrin, 1973).

a kind of embryo of the future Académie des Sciences. He also wrote various works. In the opinion of scholars, these can be divided into two periods: in the first he shows himself to be taken by a kind of religious fundamentalism; in the second, instead, he is well aware of the scientific questions.

The work in which we shall be interested belongs to the first period and indeed is the first work he wrote. It is *Quaestiones celeberrimae in Genesin* (1623).[55] In our opinion, this work, besides being a commentary to the verses of Genesis, provides the occasion for Mersenne to narrate and explain all the recent acquisitions of science in the fields of his interest. Regarding his fight against all non-Catholics, we are not able to formulate a grounded historical judgement, but one has the impression that the author is obsessed with worry about following the stances imposed by the Tridentine Council. It is sufficient to look at the title page of the *Quaestiones*.

He reaffirms his intentions in the preface (*Prefatio et Prolegomena ad Lectorem*). In fact, at the beginning, he applies to the reader (*charissime lector*) by saying that he wants to refute the atheists, the wizards, the deists and the fantastic things which they had already started to put forward all over the world (*qui iam per totum mundum grassari coeperunt*). And then: "I consider indeed a great many questions which in other respects will appear curious, so that I may show the disciples of Campanella, Bruno, Telesius, Kepler, Galileo, Gilbert, and other recent writers that their charge is false, namely, that Catholic doctors and theologians follow only Aristotle, and swear allegiance to his word, even though experiments and phenomena demonstrate the contrary".[56]

When one thinks a little about the list above, it seems a list of proscription but in that we have called the second period of his activity, he will become a friend with some of the listed scientists, particularly with Galileo.

Let us look at the text proper: it is a weighty volume which, not including the dedicatory letter, the preface and the copious indexes, consists of 1916 columns. In the same volume is also contained *Observationes et emendationes Francisci Georgi Veneti Problemata...*, which we shall not deal with. The reader's first impression is that Mersenne did not favor conciseness. Every verse is commented on through the *Problemata*, in some cases followed by the *Quaestiones* which are divided in *articuli*, in their turn divided in *capita*.

Let us begin with the comment to verse VI (*Dixit quoque Deus: fiat firmamentum in medio aquarium, et dividat aquas ab aquis*), since here Mersenne discusses the nature of the things which will be the protagonists of verse VII (the works of the second day we are interested in).

After having recalled what had happened before (that is, what was told in the preceding verses), Mersenne says that God "prepares the most adapted place for the light, precisely the firmament, which divides the lower waters from the upper and must be placed in the midst of them, but what shall we say that firmament is? What the midst? Which are the waters? What is the division? We shall say all that in the problems".[57]

As a matter of fact, in the *Problemata* we shall not find answers, but only questions. For

---

[55] Marin Mersenne, *Quaestiones celeberrimae in Genesin. Cum accurata Textus Explicatione* (Paris: Sébastien Cramoisy, 1623).

[56] Marin Mersenne, *Quaestiones*, op. cit., Prefatio, n.p.: *plurimas vero quaestiones. quae alioquin curiosae vederi possint, etiam agitasse, ut ostenderem Campanellae, Bruni, Telesij, Kepleri, Galilaei, Gilberti, & aliorum recentiorum discipulis, falsum esse, quod aiunt, Doctores, videlicet Catholicos, & Theologos solum Aristotelem sequi, & in eius verba iurare, licet experientiae, atque phaenomena contrarium evincant...* (Eng. trans. A. Williams, *The Common Expositor*, p. 179).

[57] Marin Mersenne, *Quaestiones*, op. cit., col. 800: *...locum aptissimum luci preparat, nempe firmamentum, quod inferiores aquas a superioribus dividat, atque in earum medio collocetur, sed quale dicemus esse firmamentum illud? quod medium? quas aquas? quam divisionem? in problematibus dicetur* (our Eng. trans.).

instance, as regards the firmament, in the *Problema XL*, Mersenne expounds the opinions about the firmament by many authors, each one introduced by the dubitative conjunction *an* ('or instead'), as if he expected the reader to make a choice.

We report, to substantiate what said, the beginning of the *Problema XL*: "What sort of thing is that firmament which God orders to be made on this day? Is it a certain body that is not heavy and not of great mass, but thin, loose, transparent, but nonetheless solid and resistant and such as to hold the waters? as Basil says in his third Homily, and Rupertus in Genesis I, 11, ch. 22. Or instead is it called firmament not because it is firm, but because it is the insurmountable border of the waters, as Augustine wants (*De Genesi ad litteram*, II, ch. 10)? Or instead is it called in this way because it is ingenerable and incorruptible, as Albert the Great writes (part I, *Summa de creaturis*, De IV coaequaevis)? Or instead with that word, which the Septuagint used, that is, *stereoma*, they wanted to mean that the heaven is hard and solid, as Steuchus remarks. Or instead…".[58]

But the same situation also reoccurs in the *Problema XLII* (*in medio aquarum*). To the question *Quaenam sunt illae duplices aquae?* (Which are those double waters?), he begins to answer by expounding the opinion of Basil and, after many other opinions, ends by quoting Basil and Ambrose again, without giving an opinion of his own. Finally, in the *Problema XLIII*, after having expounded the opinions of various authors, he ends by saying: "Or instead do we not know the reason why God placed the waters above the heavens? as Ascanius insinuates, after having examined all reported opinions of Fathers, to show how the Fathers can defend themselves against Peripatetics".[59]

Let us now go to verse VII (*Et fecit Deus firmamentum…*). Mersenne begins his comment in this way: "Therefore on this second day God created the firmament, between waters and waters, since it was produced from waters, since the waters are placed above, since it is transparent as the waters, therefore God placed the water above the firmament, as the phlegm in the brain is above the heart, so that the heart, the forge of heat, like a little firmament, did not burn the other parts of the microcosm with too much heat. Then God placed these waters above the firmament, in order that it proves the heavens have been created by them: as adornment of the universe; for the enjoyment of the saints existing in the empyrean heaven, so that they cheered their eyes with these crystalline and variegated waters, since the waters are suitable to assume any shape, color, ornament, as in Apocalypse 7 [Revelation 7.17] '*he will lead them to springs of living waters*' and 22 [Revelation 22.1] '*Then (the angel) showed me the river of the water of life as clear as crystal*'; and finally with his air and with his water in the empyrean heaven widely adorned of any kind of things so that the Blessed did not lack any enjoyment. Therefore waters have been divided from waters, else perhaps they would mix; they, if we trust in Hebrews, stand in this way: those which are placed above the firmament, are as far from it

---

[58] Marin Mersenne, *Quaestiones*, op. cit., col. 799: *Qualenam illud est firmamentum, quod Deus hac die fieri iubet? An illud est corpus aliquod non grave, et mole crassum, sed tenue, rarum, et percolatum, eatenus tamen solidum, et firmum quatenus aquas coerceret? ut censet D. Basil. hom. 3, et Ruper. I. 11. Gen. c.22. An propterea dicitur firmamentum, non quòd stet, sed quòd sit intransgressibilis terminus aquarum ut vult S. August. 2. De Gen. ad lit c. 10. An sic appellatur, quod sit ingenerabile, et incorruptibile? ut habet Albertus Magnus I part. summ. d e 4 coaevis q. 4. a. 19. An hoc vocabulo, quo 70 usi sunt, nempe* stereoma, *significare voluerunt coelum esse durum, et solidum, ut notat Steuchus. An…* (our Eng. trans.).

[59] Marin Mersenne, *Quaestiones*, op. cit., col. 806: *An potiùs rationem ignoramus, cur Deus super coelos aquas posverit? ut Ascanius insinuat, postquam omnes allatas Patrum opiniones examinavit, & ostendit, quomodo Patres adversus Peripateticos defendi possint* (our Eng. trans.).

as are these our lower, so that they do not touch the firmament but, to our great surprise, are suspended on high".[60]

More than a comment on verse VII, this seems a paraphrase of the verse itself. The only addition to what is said in the verse regards the nature of the firmament: Mersenne affirms that it is produced from the water and is transparent. However, this is not sufficient, and Mersenne wants to know more. The comment on verse VIII (*Vocavitque Deus firmamentum, caelum...*) allows him to investigate the nature of the firmament. In fact, at the beginning he says: "And it says, *God called firmament the heaven* which is what one sees; let us see whether the doubts which occur here can be resolved".[61]

The "doubts" are approached in the *Problema XLV*, which consists of a sequence of *an* introducing the *Quaestio VII (An firmamentum sit caelum solidum, et durum aeris, et adamantis instar; an (sicut aer) fluidum cedens, mobile, et veluti molle, ac tenuissimum)*, which regards the hardness of the firmament but also digresses about other subjects (in nine *articuli*), particularly about the comets (at that time a current topic: the observation of the motion of the comets of 1577 and of 1585 allowed Tycho Brahe to confute the immutability of the celestial spheres). Finally, in the *articulus IX (ultimam auctoris resolutionem continens)* we find two answers: "First conclusion: to me it seems not improbable that all heavens in which stars are seen to move are liquid as air";[62] ... Second conclusion: it seems more than likely that the eighth heaven, in which the stars reside, is solid".[63]

Here Mersenne identifies the firmament with the eighth heaven of the Aristotelian model and affirms that it is more than likely solid. As one can see, Mersenne does not explain himself with a definitive answer: in the first case, he uses a double negative (not improbable); in the other, "it is more than likely". As regards the first case, we have reminded above that the question had already been demonstrated by Tycho Brahe. Why this caution? The shadow of the Tridentine Council? Obviously, our study is limited to the three verses and we cannot know how Mersenne maintains the promise of the title page in the rest of the work. In our case, he seems to keep his promise: the text of the Vulgate is respected (the quotes from the Hebrew version, which we have not reported, have not left room for variations of the text) and, in the rest, nothing has been disproved.

With regard to our judgement, which may appear too laically severe, we think it proper to recall that we are talking about a work belonging to the period we have called "fundamentalist".

---

[60] Marin Mersenne, *Quaestiones*, op. cit., col. 807: *Ergo Deus hac secunda die firmamentum produxit, inter aquas, et aquas, quia ibi sunt aquae, quia ex aquis productum est, quia desuper aquae sitae sunt, quia instar aquae diaphanum est, ergo Deus super firmamentum posuit aquam, sicut phlegma in cerebro super cor, ne caloris officina, velut parvum firmamentum partes caeteras microcrismi nimio calore consumeret, ergo Deus has aquas super firmamentum collocavit, ut caelos ex eis creatos esse pateret: ad ornatum universi; ad voluptatem Sanctorum existentium in caelo empyreo, ut aquis his crystallinis, et variegatis oculos eorum recrearet, cum aquae sint omnis formae, decoris, coloris, et ornatus capacissimae, iuxta id Apocalyp. 7. Deducet eos ad vita fontes aquarum, et cap. 22, ostendit mihi fluvium aquae vivae splendidum tamquam crystallum; denique ut suus esset aer, et sua aqua caelo empyreo omni rerum specie ornatissimo, ut beati nulla voluptate careant; Ergo divisae sunt aquae ab aquis, ne forsitan confunderentur, quae, si credimus Hebraeis, ita se habent, ut quae supra firmamentum positae sunt, ab eo tantum distent, quantum hae nostrae inferiores, adeout non tangant firmamentum sed in sublimi, non sine ingenti miraculo, pendulae sint* (our Eng. trans.).

[61] Marin Mersenne, *Quaestiones*, op. cit., col. 810: *Et* vocavit, *ait* Deus firmamentum caelum *hoc quod videtur, ut autem dubia, quae hic occurrunt solvi possint, videamus* (our Eng. trans.).

[62] Marin Mersenne, *Quaestiones*, op. cit., col. 843: *Prima conclusio: caelos omnes in quibus astra moveri videntur, aeris instar liquidos esse mihi non improbabile videtur...* (our Eng. trans.).

[63] Marin Mersenne, *Quaestiones*, op. cit., col. 845: *Secunda conclusio : Probabilius esse videtur caelum octavum, in quo stellae resident, esse solidum...* (our Eng. trans.).

As scholars know, after 1630, Mersenne adhered to Copernicanism. But Arnold Williams, in his turn, was no less severe than we are: "Moreover, while the commentators strove to bring order into a confused traditional account of creation, theological irresponsibles like Paracelsus, Campanella, Bruno, and Fludd persisted in confounding the confusion. Mersenne, who wrote his commentary after all these had spun out their cosmogonies, had to include their theories in his account. In the face of such odds, the lucidity and common sense of a commentator like Pererius ought to hearten all scholars".[64]

Regarding Campanella (1568–1639), we note that he never wrote a commentary on Genesis, but he dealt with it widely in a work (written in 1616, but only published in 1622, thus a year before Mersenne's *Quaestiones*), in which he supported the ideas of Galileo about the Copernicanism.[65] As a defense of Galileo, that work might have made sense in the crucial year 1616, but that was no longer true when finally it was published in 1622. As regards the comment on the narration of the works of the first two days, it was a long exposition of the opinions of the Fathers of the Church.

---

[64] A. Williams, *The Common Expositor*, op. cit. pp. 46-47.

[65] Tommaso Capanella, *Apologia pro Galilaeo Mathematico Florentino. Ubi disquisitur, utrum ratio philosophandi quam Galilaeus celebrat, faveat sacris Scriptis an adversetur* (Frankfurt, 1622). See the English edition: Thomas Campanella, *The Defense of Galileo, A Mathematician from Florence*, Richard J. Blackwell, trans. (University of Notre Dame Press, 1994). See also the Italian edition: Tommaso Campanella *Apologia per Galileo*, Paolo Ponzio, ed. (Milan: Rusconi 1997).

Chapter 4
The Firmament and the Water Above. Fourteen Centuries of Genesis Exegeses

In the preceding chapters, we have "interviewed" more than twenty exegetes of Genesis (from Origen to Marin Mersenne) "spread" across fourteen centuries, from Origen's *Homilies* of about 240 to Mersenne's work of 1623. Our problem, now, is to understand if it is possible to derive from our interviews an interpretation that is, so to speak, historical; that is, to see whether or not the different interpretations of the famous three verses reflect what was happening in the world around, both religious and cultural, in the various periods crossed. But we must, first of all, explain a question regarding the cosmological model inherent in the text of Genesis. Though at scientific level the question has been widely explained, we think that a revisitation of the problem can result useful, since in the common opinion the conclusions of the scholars are not always shared.[1]

4.1 Moses's Cosmological Model

According to historical studies going back to the nineteenth and twentieth centuries, the final drafting of the book of Genesis, by unknown authors in Judea, dates back to the fifth-sixth centuries BC. In the field of believers, both Hebrew and Christian, the original writing of Genesis is credited to Moses at an unspecified time. The language of the original version is the Hebrew. Then, as we know, it was translated into Greek and, finally, from the Greek into Latin.

Apart from the *Homilies* of Basil the Great, whose Greek text is extant, the texts of the other exegetes we have quoted were originally in Latin, or translated from Greek into Latin (Origen). Starting with Ambrose, all exegetes made reference to the Latin text of the Vulgate. We have recalled the question of the translations, since on them is based the discussion of whether the firmament is solid or not.

Like Paul H. Seely, we too are convinced that the author(s) of Genesis wrote it while conceiving of a model of a flat earth covered by a solid dome, which was the sky with the stars. The reason why we think so is very simple. At the time of the original conception of the work, a time we can undoubtedly define as pre-scientific, the common idea regarding the earth and the sky was that corresponding to what man could intuit at first sight. To a believer, Moses (who was writing inspired by God) was obliged to remain in that order of ideas, otherwise his

---

[1]For our convenience and also to provide the reader with a reference with an extensive bibliography, we shall refer to three papers by Paul H. Seely we consider exhaustive on the question: Paul H. Seely, "The Firmament and the Water above, Part I: The Meaning of raqia in Gen 1: 6-8", *The Westminster Theological Journal* 53 (1991), pp. 227-240; "Part II: The Meaning of "The Water above the Firmament" in Gen 1 : 6-8", *WTJ* 54 (1992), pp. 31-46; "Part III: The Geographical Meaning of "Earth" and "Seas" in Gen 1: 10", *WTJ* 59 (1997), pp. 231-255. These three papers have been criticized by several people who, still nowadays, consider Genesis a textbook of astronomy. To such criticism one can answer back with the sentence quoted by Galileo in his Letter to the Grand Duchess Christina (1615) : "That the intention of the Holy Ghost is to teach us how one goes to heaven, not how heaven goes."

narration would not have been grasped by those to whom it was addressed. Hence Moses had to speak to the receivers of his message by using a language understandable in the historical context in which the message had to be read. To a non-believer, all the more reason, the author(s) of the book obviously could not think of a cosmological model different from the common one.

Therefore, in both cases (believer and non-believer), the earth which is described in Genesis is flat and the dome of the sky is solid and, as Paul Seely says: "This concept did not begin with the Greeks." He continues: "And it is precisely because ancient peoples were scientifically naive that they did not distinguish between the appearance of the sky and their scientific concept of the sky. They had no reason to doubt what their eyes told them was true, namely, that the stars above them were fixed in a solid dome and that the sky literally touched the earth at the horizon".[2]

This is what we have called Moses's cosmological model, that is, the world which is inherent in Genesis, but the exegetes we have encountered read Genesis many centuries after Greek natural philosophy had established that the earth was spherical, while maintaining the idea of the solidity of the celestial spheres in which planets and stars were fixed. Since the exegetes of the Bible belonged to a learned people, it is legitimate for us to suppose that they knew the theories (and their possible verifications) of the Greek philosophers. Let us see if and how they have faced the dilemma.

As we know, over the centuries, the exegetes of the Bible, besides interpreting it, have also theorized about how it must be interpreted, classifying the different types of exegesis. In the case of our three verses, the text has to do essentially with two terms to be interpreted: "firmament" and "waters". Therefore one can think that the subject is quite limited and the physical phenomena involved quite simple.

Leaving aside the allegorical interpretation which—let us apologize for the term—turns out to be more a literary creation than an interpretation of the text, it seems to us that the way chosen by Basil, who rejects the "shadows of high and sublime conceptions" and maintains that the water is water and not a spiritual power, will be the exegete most followed in the centuries to come. We have already spoken about the Basil's interpretation in Section 1.4, but we would like to return to the subject to insist still further on a point.

In the excerpt cited in footnote 23, Basil tells us that he is aware of all has been said about the shape of the earth by those who have written about the world and that he does not attribute less importance to Moses's narration of the creation even if Moses is silent on the shape of the earth. Being a learned man (he studied at Constantinople and at Athens), he knows the theories of Greek science, but he does not take a stand on the question. It seems, at least from his words, that, since in Genesis there is no statement about whether the earth is flat or spherical, the question remains open.

We have already pointed out how Ambrose, in his *Hexaemeron*, reverts straight to Basil's commentary, aided by his knowledge of the Greek. Perhaps, on second thought, we should not be so astonished by the fact that Lactantius maintained that the earth is flat. Even nowadays there are some people who think so! In any case, he was adhering fully to what we have called Moses's cosmological model.

At this point, it may be worthwhile to reaffirm our opinion on the science-faith conflict in the case of the interpretations of Genesis. Recalling once again that Moses's cosmological model presupposes a flat earth, in our opinion the believer can (and in any case we don't allow

---

[2] Paul H. Seely, "The Firmament and the Water Above: Part I" (1991), op. cit., p. 228.

ourselves to tell anybody what he must think) consider Genesis as a text which talks about God, and not about astronomy; the non-believer is dealing with a historical document of such importance as to have made a fundamental contribution to the foundation of Western civilization. The modest purpose of our research has been only that of following the evolution, in case there has been one, of the explanations of the "waters above the firmament", from the third to the seventeenth century, meaning with the term "evolution" the changes due to the different historical conditions.

4.2 Some remarks on the medieval annotators

Having already discussed (Section 1.6) the importance of Augustine and his ideas regarding the interpretation *ad litteram*, we now go back again to medieval annotators. All the exegetes we have considered were also believers and all, perhaps apart from Cusanus, considered Genesis by the same standards as a scientific text, even if they did not recognize this explicitly. One of them (Isidore) considered the earth flat (Section 2.1.1), the others spherical. It is not easy to grasp whether Isidore does it through ignorance or because he thinks that Moses maintains it, and, since the truth of faith must coincide with the scientific truth, the earth is flat. In any case, those who explicitly admit they know that the earth is spherical are in greater difficulty than Isidore to make the biblical text compatible with the known science.

About a century after Isidore, the personage who prevails in the field of Genesis exegesis is Bede who, however, does not show any particular innovations in comparison with Isidore, save for proposing the hypothesis of a firmament consisting of congealed water. A further confirmation of the solidity of the firmament is put forward by John Scotus, who, in his *Periphyseon* (second half of the ninth century), is the first to explicitly quote the Septuagint's Greek translation and the term *stereoma*.

As we have already recalled, John Scotus had a reputation of being very experienced in the Greek language and therefore the quotation of the meaning of the Greek term was valuable as incontrovertible evidence among his contemporaries. But Scotus's true innovation, in comparison with the preceding exegeses of Genesis, lies in his having denied the existence of true waters above the firmament. In the dialogue, he even makes use of the disciple as a stooge, making him say: "I should like it and it is necessary as well. In fact about this question, as it seems to me, no one has been convincing enough".[3] Scotus's explanation comes from his philosophy. As we shall see, Cusanus too will appeal to his philosophy (learned ignorance) to corroborate his propositions (in that case, about the earth).

We might say that, to be able to propose something new and different, one must start from a "new" philosophy. Instead Abelard, while having a new philosophical approach, limits himself to saying, "he calls firmament the airy heaven", and accounts for the possible existence of the waters above the ethereal heaven by their lightness and judges "most probable" their usefulness "for mitigating the heat of the upper fire".[4]

The medieval exegetes, before expressing their interpretations of the passages of the Bible, usually review the opinions of the *auctoritates* (the Fathers of the Church) who have preceded them. Already at the time of Abelard, the list of the *auctoritates* is quite lengthy, with Augustine always the most heeded.

In the twelfth century, according to most scholars, occurred what may be defined as "the discovery of nature", that is, the consideration of nature no longer as a thing to be

---

[3] See Section 2.1.4, footnote 41.
[4] See Section 2.2.1, footnote 48.

"contemplated", but as an object of physical research. The school which most distinguished itself for this was the School of Chartres. In our research we have been interested in two of its representatives who seemed to us the most significant, obviously from the point of view of the problem we are dealing with: William of Conches and Thierry of Chartres.

We have reported how William demolished Bede's hypothesis of the congealed waters. What is impressive is the implacable logic of the reasoning entirely based on the properties of the elements. William effectively reasons *secundum physicam,* at least to the extent that it was possible at that time. And with "in those things which concern physics, if they [the Holy Fathers] are wrong in something, it is permissible not to agree" reaffirms the freedom of thought about the things which do not "concern the Catholic faith".[5]

Here one must not misunderstand. He differs from Bede's "scientific" interpretation, nothing more: he simply gives a different interpretation of the text (there are no true waters above the firmament). Thierry too moves, also under the influence of the Platonic *Timaeus,* within the same ambit of thought of William.[6] We are also shocked by the description of the creation in the first two days: we would dare to call it a "thermodynamic" description.

In the thirteenth century, we see the flowering of some encyclopedists who renewed the deeds of Isidore but did not leave anything significant in the interpretations of the days of creation. Perhaps it is in the nature itself of the encyclopedias to supply information about what has been said, rather than to say something new.

As far as the waters above the firmament are concerned, besides the usual account of the interpretations of the preceding centuries and the denial of the existence of congealed waters, what seems to us to be a fair conclusion is that of Thomas of Cantimpré. In fact, he concludes: "But in truth, and in accordance with the Catholic doctrine, we believe in what the Church maintains, that is, that there are waters above the firmament, even if our reason cannot go so far as to grasp it".[7]

Robert Grosseteste is the philosopher who predominates in the first half of the thirteenth century. We dwelt at length on him, also quoting long excerpts of his works, on the divarication existing between the two works *Hexaemeron* and *De Luce (On Light)* in regard to the narration of the creation. In this case, in the same person we see coexisting the exegete of Genesis and the philosopher who preludes the modern science! In the second half of the same century, instead we have St. Thomas Aquinas who, apart from "giving a failing mark" to Bede's congealed waters, does not put forward new interpretations, but only provides a survey of those already proposed. Substantially, he assigns to some of them a license of legitimacy. From the point of view of an accurate exegesis, word for word, of the three verses, we can say that the medieval annotators do not differ greatly (we are obviously speaking of the interpretation *ad litteram*), except for those belonging to the School of Chartres, who try a more "physical" explanation. The true novelty which finally pushes its way in is that the "simple" language in which Genesis is written is such on purpose, to be grasped by simple people. This conclusion has the endorsement of St. Thomas and of Cusanus, who justifies the differences among the *sapientum varii conceptus* and in this conclusion is *quiescens* ("I find rest in it").[8]

---

[5] See Section 2.2.4, footnote 59.
[6] See Section 2.2.3.
[7] See Section 2.2.6, footnote 75.
[8] See Section 2.2.9, footnote 116.

4.3 The illusion of the expedient *raquia* → *expanse* in the Renaissance

The idea of considering Genesis as a text which contains an account of the creation valid at all times, also as a scientific explanation, induced the exegetes, starting at the beginning of the sixteenth century, to investigate whether the contradictions between the scientific results and the letter of the text could be solved through a different translation of the Hebraic original. According to Paul H. Seely, this occurred because "the pressure on the church from the outside to give up its belief in water above the starry firmament had become quite strong",[9] but, as we have reported above, the Tridentine Council officially established that the text of the Vulgate was the only one permitted by the Catholic Church. We have recalled many times that the Vulgate is the Latin translation of the Greek translation of the Hebraic original. Thus it is not inconceivable to suppose that a mistake may have been made in the double translation. In the sixteenth century, the sphericity of the earth was an acquired knowledge in the learned world and, if one accepted the narration of Genesis even as a "scientific" truth, the hypothesis of a mistake in the translation (which could even be imputed to the first step *raquia* → *stereoma* and then with no responsibility on part of St. Jerome) was an expedient well grounded enough to solve the contradiction of the water above the firmament.

At this point we wish to introduce the considerations we shall make in the following with an excerpt from the commentary to Genesis due to an eminent Italian biblist, Gianfranco Ravasi: "There is always the temptation to consider the biblical pages as pages which also talk about science: these pages certainly have a scientific wrapping and certainly have within them a theological core. But the wrapping is that of the science of that time. As we know, the cosmology of that time was not evolutionist: at that time any evolutionist theory whatsoever was absolutely inconceivable ... the cosmology was geocentric and visualized all the great universe centered on our little planet. It was an etiological conception (if we use the technical term) which tried to reveal the hidden causes inside the universe by using processes which were more philosophical than strictly scientific. For this reason, we have a scientific surface; there are hints connected to the science of that time, the science of the Orient; but obviously a caducous science for us moderns: even for the author itself it is only a tool for trying to elaborate his anthropological reflection, his reflection on man. In brief, we can say that this page is not a paleoanthropological page; in it the paleoanthropology can discover little, at most what was thought at that time, certainly not a precept of the Bible about science. Instead we have a precept of the Bible about man; in this sense it is an anthropological page. **And thus, we must give up the idea of using this text to try to bring it into agreement with all the present-day scientific discoveries**" (our emphasis).[10]

---

[9] P. Seely, "The Firmament and the Water Above: Part II" (1992), op. cit., n.p.

[10] *La Bibbia di Gerusalemme*, Vol. I, with the comments of Gianfranco Ravasi (Bologna: Edizioni Dehoniane, 2006), pp. 336-337: *C'è sempre la tentazione di considerare le pagine bibliche come delle pagine che parlano anche di scienza. Queste pagine hanno certamente un involucro scientifico e hanno certamente nell'interno un nucleo teologico. Ma l'involucro è quello della scienza del tempo. La cosmologia del tempo, sappiamo che era fissista: allora era assolutamente inconcepibile qualsiasi teoria evoluzionista ... la cosmologia era geocentrica e vedeva tutto il grande universo centrato sull'asse del nostro piccolo pianeta. Era una visione, come si usa dire tecnicamente eziologica, che cercava cioè di scoprire le cause segrete nell'interno dell'universo attraverso dei procedimenti più filosofici che non scientifici in senso stretto. Per cui noi abbiamo una superficie scientifica; ci sono delle indicazioni legate alla scienza di quel tempo, la scienza dell'Oriente; ma naturalmente una scienza caduca per noi moderni; anche per lo stesso autore essa è soltanto uno strumento per cercare di sviluppare la sua riflessione antropologica, la sua riflessione sull'uomo. In maniera sintetica, possiamo dire che questa pagina non è una pagina paleantropologica; in essa la pale antropologia scopre poco, al massimo scopre che cosa si pensava allora; non certo un insegnamento della Bibbia sulla scienza. Abbiamo invece un insegnamento della Bibbia*

Going back now to our subject, it is evident that all of this occurred because of the will to continue to consider Genesis as a text that also has a firm "scientific" validity and not one, as Gianfranco Ravasi says, "connected to the science of that time, the science of the Orient"; that is, one did not recognize that what we have called "Moses's cosmological model" was dated, that is, closely connected to the historical context of the time when Genesis had been conceived. In addition, Paul H. Seely reaffirms: "The historical evidence, however, which we set forth in concrete detail, shows that the *raquia* was originally conceived of as being solid and not a merely atmospheric expanse".[11] Obviously, we cannot partake in a discussion having this peculiarity (highly technical) but, to second guess, it seems to us that the solution of the "mistaken translation" was a short-lived illusion, even if it involved, besides Luther, Calvin and other Protestants, also Catholic exegetes in the sixteenth century. As we know now, the right solution would have been the simplest one: not to consider the Bible a text in which to look for scientific verities valid at any time. But this solution has difficulty in being accepted even nowadays since, right or wrong, it sparks off the science-faith conflict.

There is still a question to be asked (with regard to the subtitle of this book): why, in the case in point of the waters above the firmament, was the conflict not been sparked off by laymen? Most probably, for a long time the question did not arise for laymen, perhaps because to comment on Genesis was a duty of the churchmen and afterwards, in the sixteenth century, the Copernican question obscured any other contradiction between Genesis and the results of the natural sciences.

---

sull'uomo; in questo senso è una pagina antropologica. E allora, dobbiamo abbandonare l'idea di usare questo testo tentando di metterlo d'accordo con tutte le attuali rilevazioni scientifiche (our Eng. trans.).

[11] P. Seely, "The Firmament and the Water Above: Part I (1991)", op. cit.,, p. 227.

Author Index

A

Abelard, Peter, 50, 50n, 51, 51n, 52n, 53, 54, 56, 68, 115
Albert the Great, 65, 68,78,109
Alcuin, 42, 43, 45n
Alexander of Hales, 65
Alfraganus, 62
Alighieri, Dante, 34
Ambrose, St., 14, 15, 15n, 16, 16n, 18, 19, 23, 36, 37, 40, 68, 72, 86, 109, 113, 114
Antioch, Theophilus of, 2, 4
Aquinas, Thomas St., 46, 78, 79, 79n, 81n, 83n, 84, 91, 111n, 116
Aristotle, 3, 11, 13, 30, 44, 60, 62, 63, 65-69, 80, 93, 102, 102n, 108
Arnobius, 6
Arnoldus of Saxo, 61
Athenagoras, 2
Augustine, St., 1n, 10, 11, 14, 18,19, 19n, 20, 20n, 21n, 22, 22n, 23, 23n, 26, 29, 30, 30n, 37, 38, 40, 44, , 45, 47, 49, 52, 53, 63, 65, 68, 73, 74, 74n, 95, 100, 115
Autolycus, 4
Autun, Honorius of, 57
Auxentius, 15
Auxerre, Remigius of, 43-45
Averroes, 68

B

Barney, S. A., 34n
Bartholomew of England, 61, 63, 64
Basil the Great, 6, 10, 11-14, 40, 45, 46, 74, 79, 81, 109, 113, 114
Baur, Ludwig, 75n
Beach, J. A., 34n
Beauvais, Vincent of, 55, 68, 68n
Bede the Venerable, 38, 38n, 39, 39n, 40, 40n, 41n, 42, 45, 59, 60, 60n, 64, 68, 72, 73, 84, 115, 116
Bellarmino, Roberto, 103

Berghof, O., 34n
Bernard of Chartres, 57
Boccaletti, Dino, 69n, 87n
Bogan, M. Inez, 19n, 21n
Brahe, Tycho, 110
Brehaut, Ernest, 34, 36, 38
Bréhier, Emile, 42, 42n, 51, 51n
Bruno, Giordano, 108, 111

C

Caesarea, Eusebius of, 5, 5n, 11
Caetani, Enrico, 100
Calvin, John, 97, 97n, 98n, 99, 99n, 103, 118
Campanella, Tommaso, 108
Cantimpré, Thomas of, 61, 65, 66n, 69, 116
Carugo, Alessandro, 100n
Cassirer, Ernst, 84, 84n
Cesarini, Giuliano,85
Charlemagne, 42, 45
Charles the Bald, 45
Chartres, Bernard of, 57
Chartres, Thierry of, 56, 56n, 57, 57n, 116
Christine of Lorraine, 29
Cicero, Marcus Tullius, 8, 10, 15, 16, 19
Cipriani, Mattia, 66n
Clavius, Christopher, 99
Clement of Alexandria, 4
Cole, Henry, 90n-94n
Conches, William of, 57, 57n, 58n, 68, 116
Constant II, 15
Constantine, 6
Copernicus, Nicolaus, 10, 10n, 84, 87, 93
Cosentino, Augusto, 2n, 3n
Costabel, Pierre, 107n
Crispus, 6

D

D'Aurillac, Gerbert, 50
Dales, Richard C., 69, 69n
Danieli, Maria Ignazia, 5n
De Ferrari, Roy J., 54n, 55n

Printed in the United States
by Baker & Taylor Publisher Services